EXPERT TRADING SYSTEMS

WILEY TRADING ADVANTAGE

Trading without Fear / Richard W. Arms, Jr.
Neural Network Time Series Forecasting of Financial Markets / E. Michael Azoff
Option Market Making / Alan J. Baird
Money Management Strategies for Futures Traders / Nauzer J. Balsara
Genetic Algorithms and Investment Strategies / Richard J. Bauer, Jr.
Technical Market Indicators / Richard J. Bauer, Jr. and Julie R. Dahlquist
Seasonality / Jake Bernstein
The Hedge Fund Edge / Mark Boucher
Encyclopedia of Chart Patterns / Thomas Bulkowski
Macro Trading and Investment Strategies / Gabriel Burstein
Managed Futures / Beverly Chandler
Beyond Technical Analysis / Tushar Chande
The New Technical Trader / Tushar Chande and Stanley S. Kroll
Trading on the Edge / Guido J. Deboeck
Trading the Plan / Robert Deel
New Market Timing Techniques / Thomas R. DeMark
The New Science of Technical Analysis / Thomas R. DeMark
Point and Figure Charting / Thomas J. Dorsey
Trading for a Living / Dr. Alexander Elder
Study Guide for Trading for a Living / Dr. Alexander Elder
The Day Trader's Manual / William F. Eng
The Options Course / George A. Fontanills
The Options Course Workbook / George A. Fontanills
Trading 101 / Sunny J. Harris
Trading 102 / Sunny J. Harris
Analyzing and Forecasting Futures Prices / Anthony F. Herbst
Technical Analysis of the Options Markets / Richard Hexton
Pattern, Price & Time / James A. Hyerczyk
Profits from Natural Resources / Roland A. Jansen
The Trading Game / Ryan Jones
Trading Systems & Methods, Third Edition / Perry Kaufman
Trading to Win / Ari Kiev, M.D.
Understanding Options / Robert Kolb
The Intuitive Trader / Robert Koppel
Nonlinear Pricing / Christopher T. May
McMillan on Options / Lawrence G. McMillan
Trading on Expectations / Brendan Moynihan
Intermarket Technical Analysis / John J. Murphy
The Visual Investor / John J. Murphy
Beyond Candlesticks / Steve Nison
Forecasting Financial Markets, 3rd Edition / Mark J. Powers and Mark G. Castelino
Neural Networks in the Capital Markets / Paul Refenes
Cybernetic Trading Strategies / Murray A. Ruggiero, Jr.
The Option Advisor / Bernie G. Schaeffer
Complete Guide to the Futures Markets / Jack Schwager
Fundamental Analysis / Jack Schwager
Study Guide to Accompany Fundamental Analysis / Jack Schwager
Managed Trading / Jack Schwager
The New Market Wizards / Jack Schwager
Technical Analysis / Jack Schwager
Study Guide to Accompany Technical Analysis / Jack Schwager
Schwager on Futures / Jack Schwager
Gaming the Market / Ronald B. Shelton
The Dynamic Option Selection System / Howard L. Simons
Option Strategies, 2nd Edition / Courtney Smith
Trader Vic II / Victor Sperandeo
Campaign Trading / John Sweeney
The Trader's Tax Survival Guide, Revised Edition / Ted Tesser
The Mathematics of Money Management / Ralph Vince
Portfolio Management Formulas / Ralph Vince
The New Money Management / Ralph Vince
Trading Applications of Japanese Candlestick Charting / Gary Wagner and Brad Matheny
Trading Chaos / Bill Williams
New Trading Dimensions / Bill Williams
Long-Term Secrets to Short-Term Trading / Larry Williams
Expert Trading Systems / John R. Wolberg

EXPERT TRADING SYSTEMS

SYSTEMS

Modeling Financial Markets with Kernel Regression

JOHN R. WOLBERG

JOHN WILEY & SONS, INC.

New York • Chichester • Weinheim • Brisbane • Singapore • Toronto

Published by John Wiley & Sons, Inc.

Published simultaneously in Canada.

This publication is designed to provide accurate and authoritative information in
regard to the subject matter covered. It is sold with the understanding that the
publisher is not engaged in rendering professional services. If professional advice or
other expert assistance is required, the services of a competent professional person
should be sought.

Library of Congress Cataloging-in-Publication Data:

Wolberg, John R.
    Expert trading systems : modeling financial markets with kernel
regression / John R. Wolberg.
        p.   cm.—(Wiley trading advantage)
    Includes bibliographical references and index.
    ISBN 0-471-34508-3 (alk. paper)
    1. Capital market—Mathematical models.   2. Expert systems.
3. Regression analysis.   I. Title.   II. Series.
HG4523.W65   2000
332'.0414'015118—dc21   99-15788

Printed in the United States of America

10   9   8   7   6   5   4   3   2   1

This book is dedicated to the new generation:
Yoni and Maya Sassoon, Noa Wolberg, Abigail Kimche,
Shelly Wolberg, Aviv Kimche, and those yet to come.

סוף מעשה במחשבה תחילה.

The Result of the deed comes from the thought at the beginning.

Lecha Dodi, Shlomo Halevy Elkavatz, Circa 1500

CONTENTS

PREFACE

The use of computers for modeling financial markets has been growing exponentially. Huge sums of money are being managed based purely upon computer models. Data is fed into the models and directives are issued as output. The directives advise the user regarding suggested changes in the positions of financial instruments (e.g., stocks, bonds and commodities). The directives are based upon the results of models that predict future price or volatility changes.

This book is directed towards people in the financial community who are involved in evaluating, developing, or using market models and trading systems. Kernel regression (KR) is a modeling technique that is particularly attractive for financial market applications. Although the primary emphasis of this book is on financial modeling, the methodology can easily be applied to any high dimensional modeling problem. KR has many attractive features that make it a modeling technique that should be in the arsenal of any data mining software suite.

Attempts to model time series date back to the 1920's when Yule developed the autoregressive technique for predicting the annual number of sunspots. Since then the general subject of time series analysis has become well established. There are a number of techniques available for modeling time series in general and financial markets in particular. Financial market modeling is typically associated with large amounts of high dimensional multivariate data. Furthermore, the data typically has a low signal to noise ratio and the signals are usually nonlinear. These problems make financial market modeling particularly challenging.

Another major problem associated with financial market modeling is that one really doesn't know which if any related time series are relevant. For example, assume we are trying to develop a model to predict changes in the S&P price index. One could list a variety of series that might affect the S&P index (e.g., short and long term interest rates, commodity prices, etc.). For each related series a number of different indicators can be proposed as candidate predictors for the S&P model (e.g., the one-day fractional changes in the short term interest rate). Other candidate predictors can be suggested based upon several related series (e.g., changes in the ratio of long term to short term interest rates). One can easily develop a set of hundreds of candidate predictors that might or might not be included in the resulting model (or models).

To develop models for financial markets, methodologies are required which allow rapid analysis even though the number of candidate predictors is large. If we start with several hundred candidate predictors we will certainly eliminate most of these prior to completing the modeling task. The methodology should be geared towards finding a subset or subsets of the candidate predictor space that have an acceptable level of predictive power. KR, if applied properly, is an excellent method for developing such models. It can be extremely fast, a basic requirement when one considers the number of subsets of the candidate predictor space which might be examined as part of the modeling process. KR can be used as a stand-alone modeling technique or as a preprocessor for slower techniques such as Neural Networks. (A comparison of KR and Neural Networks is included in Appendix D. The results illustrate the complementary nature of these two modeling methodologies.)

This book is geared to three types of readers. The first group includes those who are interested in modeling in general and desire an overview of the KR technique. The second group includes those involved in the development and/or usage of KR software. The third group includes readers primarily interested in the development of computerized trading systems. The first two chapters include an introduction to the subject and an overview of the modeling process. The next three chapters include the technical details of the KR method. In Chapter 3 the mathematical basis of the KR method is developed. Chapter 4 introduces a data structure that permits an efficient implementation of KR. Chapter 5 provides information regarding the effect of the various parameters upon performance. This chapter is particularly useful for users of KR software. The methodology used in Chapter 5 utilizes artificial data and introduces a general approach to evaluating modeling software that can also be applied to other modeling techniques. Chapter 6 discusses modeling strategies and is relevant to analysts and managers interested in planning a modeling project. Some of the ideas introduced in Chapter 6 can also be applied to other modeling techniques. The final chapter discusses the application of predictions from the computer models to the development of trading systems.

The technical chapters are written for readers who are familiar with college level mathematics but are not necessarily mathematical statisticians. Many of the books on time series are directed towards the statistical community and as a result, the mathematics and notation schemes are difficult to follow for the non-statistician. The emphasis in this book is on application, evaluation, and implementation rather than topics of concern primarily to statisticians. Numerical examples rather than theorems and lemmas are used to help the reader understand the details.

John R. Wolberg
Technion - Israel Institute of Technology
Haifa, Isreal
wolberg@hitech.technion.ac.il
October, 1999

ACKNOWLEDGMENTS

I am most indebted to David Aronson for introducing me to the field of financial market forecasting. What started out as a summer project at Raden Research in the early 1980's has turned out to be a long-term relationship. Through the many projects that we've worked on together I've developed an understanding of what is required to develop models and how one goes about testing and applying them.

A number of Raden customers have given me the opportunity to develop software and to learn what the real world is all about. In particular I would like to thank Alan Bechter, Peter Borish, Ed Bosarge, John Deuss, David Hirschfeld, John Huth, Barry Honig, Paul Tudor Jones, Sandor Strauss, and Tom Wright.

Some of my colleagues at the Technion have been particularly helpful. On a number of occasions I've received statistical advice from Paul Feigin. Miriam Zacksenhouse helped me with the neural network research that forms the basis of Appendix D. Eyal Zussman was also helpful in this area. Alon Itai introduced me to the literature on the K Nearest Neighbor problem. I would also like to acknowledge the help that I have received from the many students who have taken my graduate course in The Design and Analysis of Experiments. For many years part of the course has been devoted to nonparametric statistical modeling (including kernel regression) and the feedback I've received from the students has helped me gain insight into the problem.

For software support I have relied upon Victor Leikehman and Ronen Kimche from Insightware Ltd. They are both highly talented software specialists and their help has been most appreciated. Whenever I ran into a particularly nasty bug I could always count upon either Victor or Ronen finding the source of the trouble.

To write this book I had to learn Microsoft Word and my daughter Tamar Kimche got me over the initial problems. There are still a few thousand features that I haven't learned yet but at this point I can manage!

A few people do not fall into any of the above categories but nevertheless deserve acknowledgement. I would like to thank George Butler and Richard Waldstein for their help in my early days in the modeling business. Most recently Tim Masters has been extremely helpful in answering my many questions regarding neural networks. He has published a number of books on the subject and is a real font of knowledge.

The BARRA Corporation provided the historical data used in the stock selection case study discussed in Chapter 7. I would like to thank them for being so cooperative and for their permission to include the results of the study in this book.

I would also like to acknowledge the help I received from the Technion—Israel Institute of Technology. The Technion Fund for Promotion of Research supported some aspects of the research upon which this book is based.

Finally, since I've already mentioned my daughter Tammy, I would also like to acknowledge the other members of my family just for being there and making life so pleasant: my wife Laurie, my sons Dave and Danny, my other daughter Beth, all their spouses and the grandchildren.

EXPERT TRADING SYSTEMS

1

INTRODUCTION

1.1 DATA MODELING

The concept of a mathematical model has been with us for thousands of years, going back to the ancient Egyptians and Greeks who were known for their mathematical ability. They used mathematical expressions to quantify physical ideas. For example, we are all familiar with Pythagorus's theorem:

$$C = \sqrt{A^2 + B^2} \tag{1.1}$$

which is used to compute the length of the hypotenuse of a right triangle once the lengths of the other two sides are known. For this example, C is called the dependent variable and A and B are the independent variables. To generalize the concept of a mathematical model, let us use the notation:

$$Y = f(X) \tag{1.2}$$

where Y is the dependent variable and X is the independent variable. For the more general case, both X and Y might be vectors. The function f might be expressed as a mathematical equation, or it might simply represent a *surface* that describes the relationship between the dependent and independent variables.

Data modeling is a process in which data is used to determine a mathematical model. Although there are many data

1

modeling techniques, all of them can be classified into two very broad categories: parametric and nonparametric methods. Parametric methods are those techniques that start from a known functional form for $f(X)$. Probably the most well-known parametric method is the *method of least squares*. Most books on numerical analysis include a discussion of least squares, but usually the discussion is limited to linear least squares. The more general nonlinear theory is extremely powerful and is an excellent modeling technique if $f(X)$ has a known functional form. For many problems in science and engineering, $f(X)$ is known or can be postulated based on theoretical considerations. For such cases, the task of the data modeling process is to determine the unknown parameters of $f(X)$ and perhaps some measure of their uncertainties.

As an example of this process, consider the following experiment: the count rate of a radioactive isotope is measured as a function of time. Equation (1.2) for this experiment can be expressed as

$$Y = A\, e^{-(kx)} + B \qquad\qquad (1.3)$$

where the dependent variable Y is the count rate (i.e., counts per unit time), A is the amplitude of the count rate originating from the isotope (i.e., counts per unit time at x equal to zero), k is the unknown decay constant, x is the independent variable (which for this experiment is time), and B is the background count rate. This is a very straightforward experiment, and nonlinear least squares can be used to determine the values of A, k, and B that best fit the experimental data. As a further bonus of this modeling technique, uncertainty estimates of the unknown parameters are also determined as part of the analysis.

In some problems, however, $f(X)$ is not known. For example, let's assume we wish to develop a mathematical model that gives the probability of rain tomorrow. We can propose a list of potential predictors (i.e., elements of a vector X) that can be used in the model, but there is really no known functional form for $f(X)$. Currently, a tremendous amount of interest has been generated in weather forecasting, but the general approach is to develop computer models based on data that yield predictions but are not based on simple analytical functional forms for $f(X)$.

Another area in which known functional forms for $f(X)$ are not practical are for financial markets. It would be lovely to discover a simple equation to predict the price of gold tomorrow or next week, but so far no one has ever successfully accomplished this task (or if they have, they are not talking about it). Nevertheless, billions of dollars are invested daily based on computer models for predicting movement in the financial markets. The modeling techniques for such problems are typically nonparametric and are sometimes referred to as *data-driven* methods. Within this broad class, two subclasses can be identified that cover most of the nonparametric analyses being performed today: neural networks and nonparametric regression.

The emphasis in this book is on nonparametric methods and in particular on the nonparametric kernel regression method. One problem with this class of methods is that they can be very computer intensive. Emphasis will therefore be placed on developing very efficient algorithms for data modeling using kernel regression. We live in a complicated nonlinear world, and it is often necessary to use multiple dimensions to develop a model with a reasonable degree of predictive power. Thus our discussion must include the development of multidimensional models. We must also consider how one goes about evaluating a model. Can it be used for predicting values of Y? How good are the predictions? These are the sorts of questions considered in the following chapters.

1.2 THE HILLS OF THE GALILEE PROBLEM

From my office here in Haifa, I can look out the window and see the hills of the Galilee. I've used these hills to pose a problem to students: design a program that estimates height as a function of position for any point in the Galilee. Assume that we are limited to 10,000 sample data points. Let's say we limit the problem to a square that starts from a point in Haifa port and extends 30 kilometers east and 30 kilometers north. Anyone who has seen this part of the globe knows that the Galilee is quite irregular. It includes valleys, small hills, and some larger hills that might even be called mountains.

The first simple-minded approach is to fit a grid over the entire area and distribute the data points evenly throughout this area. Spreading 10,000 points over 900 square kilometers implies a separation distance of 300 meters. This might be reasonable if we were trying to develop a model for elevation in the middle of Kansas, and it might also be reasonable for some of the valleys of the Galilee, but it is certainly not a reasonable mesh size for some of the more mountainous areas. Ideally, we would like to concentrate our points in the hillier regions and use fewer points in the flatter regions. But by doing this we introduce a new level of complexity: how do we estimate height as a function of position for a nonuniform grid?

The uniform grid suffers from a lack of resolution but allows the user an incredibly simple data structure: a two-dimensional matrix of heights. Thus, to find the height at point (X, Y), all we need to do is calculate where this point falls in the matrix. For example, consider the point $X = 12342$ and $Y = 18492$ where X is the distance in meters going east from our 0,0 point and Y is the distance in the northward direction. If we denote a point in the matrix as (I, J) and the matrix as H, then (X, Y) is located northeast of point $I = 41$, $J = 61$. We can then use simple two-dimensional linear interpolation to determine the height at (X, Y) using the four surrounding points:

$$\text{HEIGHT}(X, Y) = \sum_{i=0}^{1} \sum_{j=0}^{1} f(I + i, J + j) \qquad (1.4)$$

where

$$f(n, m) = H(n, m) * W(X, Y, n, m) \qquad (1.5)$$

and the weight terms $W(X, Y, n, m)$ are calculated as follows:

$$W(X, Y, I, J) = (300 - X + 300\,I) * (300 - Y + 300\,J)/(300*300) \quad (1.6)$$

$$W(X, Y, I + 1, J) = (X - 300\,I) * (300 - Y + 300\,J)/(300*300) \quad (1.7)$$

$$W(X, Y, I, J+1) = (300 - X + 300\,I) * (Y - 300\,J)/(300*300) \quad (1.8)$$

$$W(X,Y,I+1,J+1) = (X-300\ I) * (Y-300\ J)/(300{*}300) \qquad (1.9)$$

The problem with this simple-minded approach is illustrated in Figure 1.1. The estimate of the height of the hilltop just northeast of point (I,J) would be clearly underestimated. If the points are chosen such that we concentrate points in the hilliest regions, we introduce a number of problems:

1. If we use N points to estimate HEIGHT(X,Y), how do we choose the N points?
2. How many points should we use (i.e., what should be the value of N)?
3. Once we have found N points, what can we do to estimate the height at X,Y?

The simple data structure used for a uniform grid is no longer applicable. One approach is simply to list data in an array of length 10,000 and width 3 (i.e., column 1 is $X(i)$, column 2 is $Y(i)$, and column 3 is $H(i)$. If our choice of N points are the N nearest neighbors to (X,Y), then we will have to run through the entire

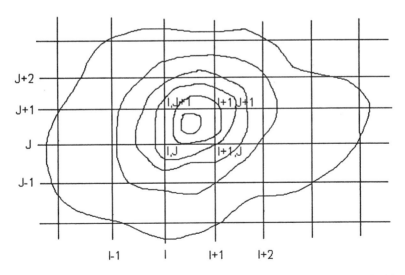

Figure 1.1 Map with lines of constant elevation. Note hilltop surrounded by four grid points.

list to find these nearest points. This would require 10,000 calculations of distance to identify the N nearest neighbors.

We could improve our search using the same 10,000 by 3 matrix by just sorting the data on one of the columns (either the X or Y column). Then to find the N nearest neighbors we could use a binary search to find the region of the matrix in which to concentrate our search. For example, let's say that the value $X = 12342$ falls between rows 4173 and 4174 of the H matrix (after sorting in the X direction). In mathematical terms, the following inequalities are satisfied:

$$A(4173, 1) <\, = X \text{ and } A(4174, 1) > X \qquad (1.10)$$

where column 1 of the A matrix is the X values of each point. Let's say that we use the four closest points for our estimation of HEIGHT(X,Y) (i.e., $N = 4$); then we need some rule to decide which rows of the matrix should be considered. One simple heuristic is to consider the 10 points below X and the 10 points above X (i.e., rows 4164 to 4183). The distance squared ($Dsqr$) from row i to the point X,Y is simply:

$$Dsqr = (X - A(i,\, 1))^2 + (Y - A(i,2))^2 \qquad (1.11)$$

We use $Dsqr$ rather than distance to avoid the need to take unnecessary square roots. The 20 values of $Dsqr$ are compared, and the points represented by the smallest four are chosen. It is possible that some or all of the four nearest points to (X,Y) fall outside this range of rows. However, more than likely we will find some, if not all, of the four nearest neighbors using this method. By sorting the data, we reduce the number of values of $Dsqr$ that must be determined from 10,000 to 20, which represents a tremendous saving in compute time (if we plan to use the program to determine elevation at many points).

Compared to the uniform grid, this choice of data structure results in a factor of 3 increase in the size of the matrix (i.e., 30,000 numbers instead of only 10,000). We also require an increase in computational time because we must now initiate a search to find the N nearest neighbors. This is the price we must pay to improve the accuracy of the computed values of elevation while limiting the size of the data array to 10,000 points.

Once the N nearest points have been located, the next problem to be addressed is the task of converting the heights of these points into an estimate of HEIGHT (X,Y). A number of approaches can be used to solve this problem, and they are discussed in later chapters. The main purpose of this discussion is to introduce the concept of estimating the value of a dependent variable (in this case HEIGHT) as a function of one or several independent variables (in this case X and Y) when the relationship is complex. Trying to determine one global equation that relates HEIGHT as a function of X and Y is useless. The best one could do is to find separate equations for many subregions in the 30 by 30 kilometer space. Using a parametric technique such as the method of least squares, an equation for HEIGHT as a function of X and Y within each region could be determined. An alternative approach is to use a nonparametric method such as kernel regression. This method is developed and discussed in the following chapters.

1.3 MODELING FINANCIAL MARKETS

The purpose of developing models for financial markets is to end up with a means for making market predictions. One typically attempts to develop models for predicting price changes or market volatility. The hope associated with such efforts is to use the model (or models) as the basis for a computerized trading system. To minimize equity drawdowns, most computerized trading systems use separate models for different markets and perhaps for different trading frequencies. By trading several models simultaneously, the equity decreases due to one model are hopefully balanced by equity increases from other models. Thus one would expect a smoother portfolio equity curve than what one might get by trading a single model. This well-known concept is called *diversification* and is discussed in most books on finance.[1] Data is fed into the models, and the system issues trading directives. The directives are usually in the form of buy and sell signals.

Using past history, one can simulate the performance of a system based on the issued signals. Thus for each model an equity curve can be generated for the simulated time period. One can look at the equity curves of each model separately and develop strategies for combining the various models into a multimodel trading system with a single combined equity curve. Whether one looks at the individual equity curves or the combined equity curve, measures of performance are required. Obviously, one must look at profitability (for example, annual rate of return), but it should be emphasized that profitability is not the sole measure of the value of a trading system. Typically, one also computes the *risk* associated with the system and then combines profitability and risk into some measure of performance. There are many definitions of risk. In his Nobel Prize-winning work on Modern Portfolio Theory, Markowitz used the standard deviations of equity changes as the measure of risk associated with an equity curve.[2] Many other definitions of risk are in use, some of which are included in a discussion of *measures of performance* in Section 1.4.

One danger associated with the modeling process described here is that if enough combinations of models and markets are tried, we will end up with a "successful" combination that only works for the modeling data but is not based on models with real predictive power. For such cases we can expect failure when the models are applied to unseen data. To protect against this possibility, one should test the entire system using unseen data (i.e., data not used in the modeling process). Only if it performs well on this data should one actually start using it to trade real funds.

The task of developing a model for a financial market is quite different from the Hills of the Galilee problem discussed in the previous section. That problem exhibited the following characteristics:

1. The number of independent variables (i.e., 2) was known.
2. For every combination of the independent variables (i.e., X and Y), there was one correct value of the dependent variable (i.e., H).
3. The dependent variable H could be modeled as a function of X and Y, and with enough data points the model could be made to be as accurate as desired.

The modeling of financial markets is quite different in all of these respects.

1. The number of independent variables required to develop a model with a reasonable degree of predictive power is unknown. Indeed, one does not even know if it is possible to obtain a decent model with the available data.

2. For a given set of independent variables, there is no guarantee that if the values of the independent variables are the same for two data points, the values of the dependent variable will also be the same. In other words, for each combination of the independent variables there is a range of possible values of the dependent variable.

3. Regardless of the number of available data points, there is no hope of converging to a model free of error. (In other words, we assume that our final model will contain some noise. Our hope is to obtain a model in which the signal is strong enough that the predictive power of the model will be of some value.)

Financial markets can be characterized as having a low signal-to-noise ratio. In other words, a large fraction of the change in price from one time period to the next appears to be a *random shock*. In addition, the small signal typically varies in a highly nonlinear manner over the modeling space. Often, however, the random shock is not totally random if one considers other related time series. By bringing in more relevant information (i.e., related time series), a greater fraction of the price changes can be explained. The interesting aspect of financial market modeling is that one does not need to obtain a high degree of predictability to develop a successful trading system. For example, if models could be developed for a dozen different markets and each model could consistently explain 5 percent of the variance in the price changes (see Section 1.4 for a definition of Variance Reduction), a highly profitable trading system could be developed based on these models!

As an example of the modeling process, consider the problem of predicting a future price of gold. What are the independent variables? The best that an analyst can do is to propose a set of candidate predictors. Each of these predictors must be *backward looking*. In other words, their values must be known at the times the predictions are made. How does one come up with a decent set of candidate predictors? A massive body of research has been devoted to this problem. Many articles and books have been written on this subject for many different financial markets. This book concentrates on how to select a model once a set of candidate predictors has been proposed. The focus of the book is on evaluating the candidate predictors individually and together once they have been specified. However, some comments are made in Chapter 2 regarding the specification of candidate predictors.

The number of candidate predictors available for developing prediction models is limited only by the imagination of the analyst. The primary source of candidate predictors is from the time series being modeled. For the gold model, one would first use the gold price series as a source of candidate predictors. Often a number of predictors might be variations on the same theme. For the gold example, the most obvious choices of candidate predictors are past changes in gold prices. For example, the relative change in the prices of gold over one, two, and three time periods can be selected as the first three candidate predictors:

1. $X1 = 1 - \text{LAG(GOLD,1)/GOLD}$
2. $X2 = 1 - \text{LAG(GOLD,2)/GOLD}$
3. $X3 = 1 - \text{LAG(GOLD,3)/GOLD}$

The variable GOLD represents a time series of gold prices, and the LAG operator returns the series being lagged by the number of records indicated as the second parameter. The next theme might be ratios based on the current price of gold and moving averages of gold prices. For example:

1. $X4 = \text{GOLD/MA(GOLD,3)} - 1$
2. $X5 = \text{GOLD/MA(GOLD,10)} - 1$
3. $X6 = \text{GOLD/MA(GOLD,50)} - 1$

The operator MA is a moving average over the number of time periods indicated by the second parameter. The candidate predictor $X4$ is the deviation from 1 of the ratio of the current price of gold divided by the moving average of gold over the last three time periods. The candidate predictor $X5$ is similar to $X4$ but is based on a longer time period, and $X6$ is based on the longest time period. Negative values of these three candidate predictors mean that the latest price of gold is less than the three moving averages. Next, we might start considering data from other markets: (e.g., $X7 = 1 - \text{LAG}(S\&P, 1)/S\&P$). The variable S&P represents a time series of the S&P price index. When one starts considering all the possible predictors that might influence the future change in the price of gold, the set of candidate predictors can become huge.

A reasonable approach to modeling when the number of candidate predictors is large is to consider subspaces. For example, assume we have 100 candidate predictors. We might first consider all 1D (one-dimensional) spaces. We try to find a model based on $X1$ (i.e., $Y = f(X1)$), then $f(X2)$, up to $f(X100)$. After all 1D spaces have been considered, we then proceed to 2D spaces. If we examine all 2D combinations (i.e., $f(X1,X2)$ up to $f(X99, X100)$), then we must consider $100*99/2 = 4950$ different 2D spaces. Proceeding to 3D spaces, the number of combinations increases dramatically. For 100 candidate predictors, the total number of 3D spaces is $100*99*98/6 = 161700$. Clearly, some sort of strategy must be selected that limits the process to an examination of only the spaces that offer the greatest probability of success.

One question that comes to mind is the upper limit for the dimensionality of the model. How far should the process be continued? three dimensions? four dimensions? For a given number of data points, as the dimensionality of the model increases, the sparseness of the data also increases. To discuss sparseness from a more quantitative point of view, let's first define a *region* in a space as a portion of the space in which all the signs of the values of the various X's comprising the space do not change. For example, if we have N data points spread out in a 2D space, then we have an average of $N/4$ points per region. Let's say we are looking at the 2D region made up of candidate predictors $X5$ and $X17$. Assume further that both $X5$ and $X17$ have been normalized so that their means are zero. The four regions are:

1. $X5 > 0$ and $X17 > 0$
2. $X5 < 0$ and $X17 > 0$
3. $X5 > 0$ and $X17 < 0$
4. $X5 < 0$ and $X17 < 0$

Now if we spread the same N points throughout a 3D space, we have eight separate regions and the average number of points per region is reduced to $N/8$. Generalizing this concept to a d dimensional space, the number of points per region is $N/2^d$. (In other words, for every increase by one in the dimensionality of the model, the average number of data points per region is halved.) Thus we can conclude that the value of N (the number of available data points for development of the model) can be used to set the maximum dimensionality of the model. All we have to do is to set a minimum value for the average number of data points per region. Let's say we have 1000 points and we want at least an average value of 10 points per region. The maximum dimensionality of the model would thus be determined by setting $1000/2^d$ to 10, which leads to a value of $dmax$ equal to 6. (Increasing d to 7 leads to an average number of data points per region of $1000/128 < 10$.) If the distribution of points in a particular dimension is not normal, then the definition of $region$ must be modified. Nevertheless, the conclusion is the same: With increasing dimensionality, an exponentially increasing N is needed to maintain the data density. It should be mentioned that certain modeling strategies allow a greater number of variables to be included within the model without causing excessive sparseness. For example, the methods of principal components and factor analysis are well-known techniques for addressing this problem. Another approach is to use multistage modeling, a technique discussed in Chapter 4.

In summary, the process of modeling financial markets can be described as follows:

1. Specify a list of candidate predictors (X's) and gather the appropriate data to compute the X's for the time period that is proposed for modeling. It is not necessary to require that all data be in one concurrent time period.

2. For the same time period (or periods), compute the values of the dependent variable Y (the quantity to be modeled).

3. Determine a maximum value for the dimensionality of the model.

4. Specify a criterion (or criteria) for evaluating a particular model space. Some well-known criteria are discussed in the next section.

5. Specify a strategy for exploring the spaces.

6. For each space compute the value of the criterion (or criteria) used to select the "best" model (or models).

7. If there is sufficient data, test the "best" models using this remaining unseen data (i.e., data that has not yet been used in the modeling process).

There are many details that are not included in this list. However, this list covers the general concepts required to model financial markets. Specifics regarding these various stages in the modeling process are considered in later chapters.

1.4 EVALUATING A MODEL

Typically, when a nonparametric method is used for data modeling, a variety of models are proposed and then some sort of procedure must be used that permits selection of the *best* model. When there are a number of candidate predictors, combinations of the predictors are often examined to see which *space* is best. In this section several popular definitions of *best* model are considered.

A number of different strategies may be used to evaluate a model. If sufficient data are available, the usual choice is to use some of the data as *learning* data (i.e., to generate or train the model) and some of the data for *testing*. Since one typically looks at many different models in the search for the *best* model, one question that should be asked is the following: if we finally end up with a good model, is it real or did it just happen by chance after looking at many potential models (i.e., spaces and parameters). A strategy used to answer this question is to save yet a third set of data: the *evaluation* data set. This data is used only after the modeling process has been completed. The final model is applied to this data to see if the model succeeds for unseen data.

What do we mean by the *best* model? The definition of *best* is, of course, problem dependent. There are several different well-known definitions, but the modeling process need not be limited to these standard definitions. It is useful, however, to mention some of the most popular modeling criteria. A very useful and popular criterion for data modeling is Variance Reduction (VR). If the purpose of our model is to predict Y, once a model has been proposed, it can be used to predict Y for a series of $ntst$ test data points. In other words, for test point i there is a known value of $Y(i)$ and a value that is calculated using the model: $Ycalc(i)$. VR is computed as follows:

$$
\text{VR} = 100 * \left(1 - \frac{\sum\limits_{i=1}^{ntst} (Y(i) - Ycalc(i))^2}{\sum\limits_{i=1}^{ntst} (Y(i) - Yavg)^2} \right) \tag{1.12}
$$

In this equation, *Yavg* is the average value of all the values of $Y(i)$. This equation shows that VR is the percentage of the variance in the data that is explained by the model. A perfect model (i.e., a model that yields values of $Ycalc(i) = Y(i)$ for all test points) has a value of VR equal to 100 percent. A value of VR close to zero means that the model has little or no predictive power.

An interesting question is, what can we say about the expected value of VR for a model with absolutely no predictive power? Let's say we have a random time series of X and Y values. We divide the data into learning and test data sets. The values of *Yavg* for the two sets should be close, but there will be a difference. Any modeling technique that attempts to calculate values of *Ycalc* for the test data based on the learning set data will be slightly biased due to this slight difference in the means. Thus one can expect a slightly negative VR rather than a value of zero. (This point is treated in greater detail in Appendix B.)

Another criterion often used in data modeling is the root mean square error (RMSE). The computation of RMSE is similar to VR:

$$\text{RMSE} = \sqrt{\sum (Y(i) - Ycalc(i))^2 / ntst} \qquad (1.13)$$

The problem with RMSE is that it has the dimensions of Y whereas VR is a dimensionless quantity. For example, if two models are compared, one for predicting changes in the price of gold and the other for predicting changes in the price of silver, a comparison of the two values of RMSE will be meaningless. These RMSE values would have to be normalized in some way that would introduce the relative prices of gold and silver. Alternatively, the comparison of the two values of VR gives a direct indication regarding the relative worth of the two models.

Both VR and RMSE suffer from the problem of outliers. That is, if the data include some points that are far from the model surface, then these points tend to have a disproportionate effect on the evaluation of the model. To reduce the effect of outliers, a number of "robust" modeling criteria have been proposed. Probably the most popular criterion in this category is the root median square error (RMedSE):

$$\text{RmedSE} = \sqrt{med\big((Y(i) - Ycalc(i))^2\big)} \qquad (1.14)$$

where *med* is the median operator. To turn RmedSE into a dimensionless variable similar to VR, the median variance reduction (MVR) can be defined as follows:

$$\text{MVR} = 100* \, (1 - med((Y(i) - Ycalc(i))^2)/ \\ med((Y(i) - Yavg)^2)) \qquad (1.15)$$

Robust criteria provide a straightforward answer to the problem of outliers, but they add to the computational complexity of the modeling process. The computation *of* RMedSE and MVR requires a sort of the values of $(Y(i) - Ycalc(i))^2$. Sorting typically requires times of order $N*log2(N)$. Depending on N (the number of test points), this can be an important factor in determining the compute time required for the modeling process. An additional problem is that the values of $Ycalc(i)$ must be saved in order to compute RMedSE or MVR. Alternatively, VR and RMSE can be determined at almost no extra cost, and they do not even require saving the values of $Ycalc(i)$.

Another measure of the performance of a model is Fraction Same Sign (FSS). The FSS value is the fraction of the $ntst$ predictions in which the signs of $Ycalc(i)$ and $Y(i)$. are the same. If we can effectively predict the sign of $Y(i)$, the development of a useful trading strategy is straightforward. Since the computation of FSS is so simple, it is useful to include this output parameter in any output report regardless of whether or not it is used as the modeling criterion. One problem with FSS as defined above is that values of $Ycalc(i)$ and $Y(i)$. close to zero are treated the same as values far from zero. In Section 6.6 a variation on the definition of FSS is introduced that avoids this shortcoming.

A well-known modeling criterion, correlation coefficient (CC), measures the degree of correlation between $Ycalc(i)$ and $Y(i)$. CC is defined as:

$$CC = \frac{\sum (Y(i) - Y_{avg})(Ycalc(i) - Ycalc_{avg})}{\sqrt{\sum (Y(i) - Y_{avg})^2 \sum (Ycalc(i) - Ycalc_{avg})^2}} \qquad (1.16)$$

To actually compute CC there is no need to use Eq. (1.16). This equation requires two passes through the data: the first pass is required to compute the average values, and the second pass is then required to sum the differences. A direct single pass calculation is performed as follows:

$$\text{SUMYYC} = \text{sum}(Y*Ycalc) - \text{sum}(Y^2)*\text{sum}(Ycalc^2)/ntst \qquad (1.17)$$

$$\text{SUMY2} = \text{sum}(Y^2) - (\text{sum}(Y))^2/ntst \qquad (1.18)$$

$$\text{SUMYC2} = \text{sum}(Ycalc^2) - (\text{sum}(Ycalc))^2/ntst \qquad (1.19)$$

$$\text{CC} = \text{SUMYYC}/\text{sqrt}(\text{SUMY2} * \text{SUMYC2}) \qquad (1.20)$$

The sum operator is the scalar sum of all terms of the vector argument. The squaring operator on a vector yields a vector output of the same length as the input but with each element squared.

There is an important weakness in using the correlation coefficient as a modeling criterion for financial markets. A high degree of correlation can exist, and yet the predicted values of $Y(i)$ (i.e., $Ycalc(i)$) might not be particularly useful. Consider, for example, a case in which $Ycalc(i)$ ranges between 1 and 1.2, while the values of $Y(i)$ range between −2 and 2. Even if the correlation coefficient is significant, there is no meaningful interpretation of the resulting values of $Ycalc(i)$. The model is always predicting that the value of $Y(i)$ will be positive, even though the range includes negative values. Correlation can still be used as a modeling criterion if the coefficient is defined as Correlation Coefficient through the Origin (CCO). This parameter assumes a linear relationship between $Y(i)$ and $Ycalc(i)$, but the assumption is that the relationship goes through the origin of the $Y - Ycalc$ plane. CCO is computed as follows:

$$CCO = \text{sum } (Y*Ycalc)/\text{sqrt}((\text{sum}(Y^2)*\text{sum}(Ycalc^2))) \qquad (1.21)$$

Values of CCO, like the standard correlation coefficient, range between −1 and 1. A value of 1 implies that all values fall exactly on a line in the $Y - Ycalc$ plane that goes through the 0,0 point. Furthermore, positive values of $Ycalc(i)$ correspond to positive values of $Y(i)$, and negative values correspond to negative values. The value of CCO is not related to the slope of the line.

When modeling financial markets, an alternative approach is to use a modeling criterion based on trading performance. Rather than measuring how close the values of $Ycalc(i)$ are to $Y(i)$, the values of $Ycalc(i)$ can be used to generate trades. Using the trades, one can generate an equity curve and then use this curve to evaluate the quality of the predictions. In this manner, a trading-based measure of performance can be obtained that is then used in the selection process needed to choose the best set of candidate predictors. A number of parameters must be specified to describe how the system trades based on the values of $Ycalc(i)$. Examples of such parameters are buy and sell thresholds for entering and exiting a trade. The problem with this approach is the tendency to *overfit*. By varying the various thresholds through a wide range of values, one can end up with results that are excellent in the testing period but are disappointing when used in production.

This section defines some of the better known measures of performance. It should be clear that many other measures of performance are possible. Section 6.6 considers this subject in greater detail and discusses some measures of particular application to financial market modeling.

1.5 NONPARAMETRIC METHODS

Nonparametric methods of data modeling predate the modern computer age. In the 1920s two of the giants of statistics (Sir R. A. Fisher and E. S. Pearson) debated the value of such methods.[3] Fisher correctly pointed out that a parametric approach is inherently more efficient. Pearson was also correct in stating that if the true relationship between X and Y is unknown, then an erroneous specification in the function $f(X)$ introduces a model bias that might be disastrous.

Hardle includes a number of examples of successful nonparametric models, the most impressive of which is the relationship between change in height (cm/year) and age of women (Figure 1.2).[3] A previously undetected growth spurt at around age 8 was noted when the data was modeled using a nonparametric smoother.[4] To measure such an effect using parametric techniques, one would have to anticipate this result and include a suitable term in $f(X)$.

Clearly, one can combine nonparametric and parametric modeling techniques. A possible strategy is to use nonparametric methods on an exploratory basis and then use the results to specify a parametric model. However, as the dimensionality of the model and the complexity of the surface increase, the hope of specifying a parametric model becomes more and more remote. For financial market modeling, parametric models are not really feasible. As a result, considerable interest has been shown in applying nonparametric methods to financial market modeling. Recent books on the subject include Bauer (1994),[5] Gately (1995),[6] and Refenes (1995).[7]

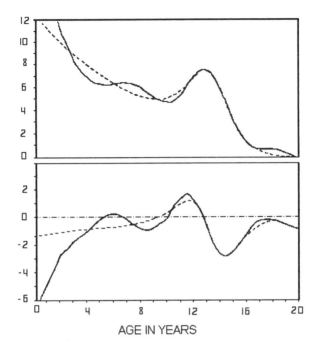

Figure 1.2   Human growth in women versus age. The top graph is in cm/year. The bottom graph is acceleration in cm/year^2. The solid lines are from a model based upon nonparametric smoothing and the dashed lines are from a parametric fit.[3,4]

The emphasis on neural networks as a nonparametric modeling tool is particularly attractive for time series modeling. The basic architecture of a typical neural network is based on interconnected elements called neurons, as shown in Figure 1.3. The input vector may include any number of variables. All the internal elements are interconnected, and the final output for a given combination of the input parameters is a predicted value of Y. Weighting coefficients are associated with the inputs to each element. If a particular interaction has no influence on the model output, the associated weight for the element should be close to zero. As new values of Y become available, they can be fed back into the network to update the weights. Thus the neural network can be *adaptive* for time series modeling: in other words, the model has the ability to change over time.

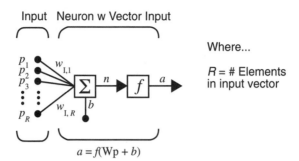

Figure 1.3 A typical element in a neural network. The Σ block sums the weighted inputs and the bias b. The f block is a nonlinear transfer function.

One major problem is associated with neural network modeling of financial markets: the huge amount of computer time required to generate a model. If one wishes to use tens or even hundreds of thousands of data records and hundreds of candidate predictors, the required computer time is monumental. To have any hope of success, techniques are required to preprocess the data in order to reduce the number of candidate predictors to a reasonable amount. The definition of *reasonable* varies, of course, depending on the available computing power. However, regardless of the hardware available, preprocessing strategies are essential to successfully apply neural networks to financial market modeling. Use of kernel regression is an alternative modeling strategy that can be many orders of magnitude faster than more computer-intensive methods such as neural networks. Kernel regression techniques can be used to very rapidly obtain the *information-rich* subsets of the total candidate predictor space. These subspaces can in turn be used as inputs to a neural network modeling program. For a comparison of neural network and kernel regression modeling, see Appendix D.

1.6 FUNDAMENTAL VERSUS TECHNICAL ANALYSIS

Chapter 1 of Jack Schwager's book *Fundamental Analysis* is entitled "The Great Fundamental versus Technical Analysis

Debate.[8] In that chapter Schwager defines fundamental analysis as analysis involving the use of economic data (e.g., production, consumption, disposable income) to forecast prices. He defines technical analysis as analysis based primarily on the study of patterns in the price data itself (and perhaps volume and open interest data).

The popularity of these differing approaches has evolved over time. In the early 1970s, most serious financial analysts regarded technical analysis with disdain. But as the decade wore on, the huge price trends that developed in the commodities markets were highly favorable for trend-following techniques. Technical analysis was ideally suited to capture these movements, and this approach became the order of the day. By the late 1980s, technical analysis was the dominant approach for making trading decisions. However, nothing lasts forever, and as Schwager noted, "the general market behavior became increasingly erratic." Choppy markets are notoriously unfriendly toward trend followers, and this had a real damping effect on technical analysis. Once again, fundamental analysis became more popular. What seems to have evolved is the tendency to combine fundamental analysis for making longer-term predictions and using technical analysis for shorter-term market timing.

The question that one might ask is, Where does the approach discussed in this book fit into the grand scheme of things? Is it technical analysis or fundamental analysis? Clearly, we are looking for patterns, so this might be construed as a technical approach. On the other hand, the analyst is encouraged to use economic data as well as basic price, volume, and open interest data. Once one has the ability to look at hundreds of candidate predictors, there is no need to limit the search for a model to the confines of the very simple price and volume-related indicators. Relevant series such as interest rates and currency exchange rates are fair game for the analyst. Perhaps one should say that the distinction between fundamental and technical analysis becomes moot when one is using a multivariate prediction approach to modeling.

In the last few years, the power of the computer has grown so enormously that the type of analysis proposed in this book has become increasingly cost effective. By using the types of algorithms described in Chapter 4, vast numbers of potential subspaces can be examined in a relatively short period of time. I recently did some consulting work in which I searched for a model using 658 candidate predictors based on 2377 data records. All of the 658 X's were first examined individually. The best 35 (on the basis of Variance Reduction) were then used to form two-dimensional spaces using all other variables. The number of 2D spaces examined was 35*34/2 + 35*(658 − 35) = 22400 (i.e., all pairs of the best 35 plus all pairs of each of the 35 with the remaining 623). The 50 best 2D spaces were then used to create 3D spaces. A grand total of 55737 spaces were examined in less than two hours using a relatively slow computer (a Pentium 100). The average time per space was about 0.1 second. Another analysis was based on a combined database of 48 stocks over 2041 days (i.e., a total of 48*2041 = 97968 data records). This data set included 23 X's, and a total of 524 spaces were examined in about 6000 seconds. Even for this huge data set, the average time per space was only about 11 seconds. These rates can already be improved by over a factor of five going to the faster processors on the market today. With symmetric multiprocessor hardware about to become standard, even greater speeds can be expected in the near future.

NOTES

1. See J. C. Francis, *Investments, Analysis and Management* (New York: McGraw-Hill, 1980).

2. H. Markowitz, "Portfolio Selection," *Journal of Finance* (March 1952).

3. W. Hardle, *Applied Nonparametric Regression* (Cambridge, UK: Cambridge University Press, 1990).

4. T. Gasser, H. G. Muller, W. Kohler, L. Molianari, and A. Prader, "Nonparametric Regression Analysis of Growth Curves," *Annals of Statistics* 12 (1984): 210–229.

5. R. J. Bauer, *Genetic Algorithms and Investment Strategies* (New York: John Wiley & Sons, 1994).

6. E. Gately, *Neural Networks for Financial Forecasting* (New York: John Wiley & Sons, 1996).

7. A. P. Refenes, *Neural Networks in the Capital Markets* (New York: John Wiley & Sons, 1995).

8. J. Schwager, *Fundamental Analysis* (New York: John Wiley & Sons, 1995).

2

DATA MODELING OF TIME SERIES

2.1 THE TIME SERIES PROBLEM

Time series present some unique modeling problems. When dealing with time series, the analyst should ask the following questions:

1. Is there enough data available to develop and test a model?
2. For the proposed candidate predictors, does the available data adequately populate the various spaces?
3. Is serial correlation a major problem?
4. Does the model hold up over time? (In other words, can we be reasonably assured that the model itself doesn't change with time? Or if it does change, is the change gradual?)

1. Is there enough data available? When modeling financial markets, the availability of data is dependent on the time scale of the data. For example, if one is using daily data, then there are only about 250 data points per year. Ten years of data would thus include about 2500 points, but then question number (4) becomes relevant: Does the market behavior 10 years ago have any relevance today? One must develop and use testing procedures to ascertain the relevance of models. If the modeling is based on a much more rapid time scale (say, five-minute time periods), then much more data is available per day. For example, for markets that open at 9:30 A.M. and close at 16 P.M., there are 78 five-minute time periods per day. So in one year there are approximately 78*250 (i.e., 19500) available data points. Clearly, it is much easier to develop models when alot more data is available.

One method used to increase the amount of available data is to include data from a group of financial series rather than just a single series. For example, consider a common stock database in which daily prices are available. Rather than trying to model the fractional price changes for one particular stock, it is much easier to model a group of similar stocks (for example, from a particular industrial group). An application in which monthly data from over 7000 different stocks are analyzed as a group is discussed in Chapters 6 and 7.

2. Does the data adequately populate the spaces under consideration? Coverage of the space is indeed an important concept in modeling. When modeling financial markets, one first must make sure that the different types of markets are adequately represented in both the data used to develop the model and the data used to test the model. For example, consider a situation in which the modeling data is taken only from time periods that can be described as "bullish" (i.e., the price changes of the financial instrument of interest are generally positive). If the testing of the model is based on data from "bearish" periods (i.e., falling prices), then one should not be surprised if the testing yields disappointing results. In Figure 2.1 the trend is essentially *bullish* up to record 20. From about record 25 the trend becomes *bearish*. It is preferable to use data that is representative of the different types of market conditions that one is likely to encounter.

The same argument holds true for the independent variables. For example, consider an indicator x, which is some sort of oscillator about zero. If the values of x are mostly positive during the modeling time periods and then mostly negative during the testing time periods, once again one might expect poor results. It is useful to devise statistical tests to measure the similarity of the modeling and testing data.

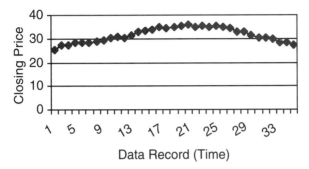

Figure 2.1 Price data with bullish and bearish periods.

3. Is serial correlation a problem? Serial correlation is a problem unique to time series. Serial correlation relates to the independence of adjoining points. Are the adjoining points independent or are they somehow related? We can define r (the correlation coefficient) between two variables (let's say x and y) as follows:

$$r = \frac{SS_{xy}}{\sqrt{SS_{xx} SS_{yy}}} \tag{2.1}$$

where

$$SS_{xy} = \sum_{i=1}^{i=n} (x_i - \bar{x}) * (y_i - \bar{y}) \tag{2.2}$$

$$SS_{xx} = \sum (x_i - \bar{x})^2 \tag{2.3}$$

$$SS_{yy} = \sum_{i=1}^{i=n} (y_i - \bar{y})^2 \tag{2.4}$$

The notation SS is used for "sum of the squares." $\bar{x}$ and $\bar{y}$ are the average values of x and y. Consider the case where x is a candidate predictor and y is the same predictor lagged by one time period. Serial correlation implies that the value of r is significantly larger than 0. To illustrate this point, assume that x and y are defined as follows:

$$X = \text{MA(SERIES,10)} \qquad (2.5)$$

$$Y = \text{LAG(MA(SERIES,10), 1)} \qquad (2.6)$$

Figure 2.2 shows 21 values of SERIES and X computed from SERIES. The serial correlation of SERIES is −0.175, which is fairly close to zero, while the serial correlation of X is 0.623.

The high degree of correlation between adjoining values of x should not be surprising. The values of y (i.e., x lagged by one time period) are determined from almost the same 10 points as the values of x. What does this imply? Since the values of x are serially correlated, we really don't have as much independent data as we might expect from just considering the number of data points. Serial correlation is not necessarily a major problem. It is merely a fact of life when modeling time series data, and it should be considered when one is developing a modeling strategy.

The Durbin–Watson test can be used to determine if the serial correlation of a time series is significant.[1] Descriptions of this test and Durbin–Watson tables are included in many texts.[2] A d statistic is computed, which ranges from 0 to 4:

$$d = \frac{\sum\limits_{t=2}^{n}(x_t - x_{t-1})^2}{\sum\limits_{i=1}^{n}x_t^{\,2}} \qquad (2.7)$$

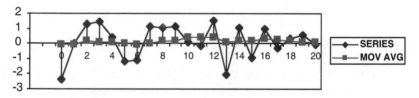

Figure 2.2 A time series and its moving average.

If the series does not exhibit serial correlation, then d is approximately 2; if the values are highly positively correlated, then d is close to 0; and if they are highly negatively correlated, then d is close to 4.

4. Does the model hold up over time? The *persistency* of models is an important concept when the models are used to make future predictions. In financial markets one can discover many examples of models that work quite well up to a certain point in time and then cease to perform acceptably. The user is then faced with trying to decide if the failure is temporary or is due to some underlying fundamental change in the market behavior. There is no simple answer to this question. However, it can be said that this problem is much more relevant for models developed using data from a longer time period. For example, models based on daily data are much more likely to become obsolete than models based on five-minute bar data. The most obvious check for any model's persistency is to save the most recent data for final testing. If the model holds up at this point, then at least one can start using it with some degree of certainty that it is still a useful tool.

2.2 CLASSICAL METHODS OF TIME SERIES MODELING

The history of time series analysis predates the computer age. H. Wold wrote a book in 1938 that summarized knowledge of the subject to that point in time.[3] Another early book on the subject was written by the well-known cyberneticist Norbert Weiner.[4] Both Wold and Weiner refer to the contributions of G. U. Yule, A. Khinchine, and A. Kolmogoroff to time series analysis. In the introduction to a fairly recent book, A. S. Weigend and N. A. Gershenfeld reflect on the early work of Yule: "The beginning of 'modern' time series prediction might be set in 1927 when Yule invented the autoregressive technique in order to predict the annual number of sunspots. His model predicted the next value as a weighted sum of previous observations of the series. For the half century following Yule, the reigning paradigm remained that of linear models driven by noise."[5]

A summary of classical time series analysis is included in the introduction of a well-known book by M. B. Priestley.

> During the past fifty years or so, time series analysis has become a highly developed subject, and there are now well-established methods for fitting a wide range of models to time series data—as described in the books by Anderson [1971], Box and Jenkins [1970], Brillinger [1975], Chatfield [1975], Hannon [1970], Jenkins and Watts [1968], Koopmans [1975] and Priestley [1981]. However, virtually all the established methods rest on two fundamental assumptions; namely that (i) the series is *stationary* (or can be reduced to stationarity by some transformation, such as differencing), and (ii) the series conforms to *a linear* model. Assumption (i) means, in effect, that the main statistical properties of the series remain constant over time, and (ii) means that the values of the observed series can be represented as linear combinations of present and past values of a "strictly random" (or "independent") series. Needless to say, both of these assumptions are mathematical idealizations which, in some cases, may be valid only as approximations to the real situation. In practical applications the most one could hope for is that, for example, over the observed time interval the series would not depart "too far" from stationarity for the results to be invalid.[6]

One of the best known texts on the subject of time series analysis is the book by Box and Jenkins.[7] The first edition of this book was written in 1969 and reflected the growing power of computers for solving time series problems. The widespread availability of computers influenced many people to apply computers to the modeling of time series data. It was obvious to most analysts that the ability to predict the future based on past history could be incredibly valuable. The techniques popularized by Box and Jenkins were applied to many areas of activity. Analysts on Wall Street were among the early users of these techniques as they attempted to make predictions related to the financial markets.

Weigend and Gershenfeld trace the first example of the recognition of the limits of linear modeling to one of the early pioneers in time series analysis, S. Ulam. In 1957 Ulam discussed the problems associated with predicting the next values of the time series generated with the following simple equation: $x_{t+1} = \lambda x_t(1 - x_t)$. Linear modeling fails for this problem. In 1980 Tong and Lim described a technique called the threshold autoregressive model (TAR) which is considered the first globally nonlinear method for modeling time series. A description of the method is included in Tong's book.[8] Since then a lot of interesting work on nonlinear time series analysis has been reported. Contributions from many of the leaders in the field are included in a book edited by T. Subba Rao.[9]

The general subject of *data mining* has received considerable attention in recent years. The combination of huge databases and very powerful computers has led to a thriving modeling industry. Many academics and industrialists have realized that benefits can be obtained if tools are available for extracting information from their databases. As a result, commercial software based upon well-known techniques is readily available. Weiss and Indurkhya list three basic categories into which most popular techniques fall:[10]

1. Techniques based upon derivation of mathematical equation (e.g., classical regression and neural networks).
2. Techniques based upon development of logical rules (e.g., CART, *www.salford-systems.com*).
3. Techniques based upon similarity (or distance) (e.g., kernel regression).

Within each category there are variations, and hybrid technologies are becoming popular. All of these techniques can be used to develop nonlinear models in which a variety of candidate predictors are considered. Access to many publicly and commercially available data mining software products can be found at the Internet site *www.kdnuggets.com*.

2.3 THE CURSE OF DIMENSIONALITY

The concept of *the curse of dimensionality* relates to all very complex systems: the amount of data required increases exponentially with the number of variables appearing in the model.[11] The greater the system complexity, the greater the number of dimensions required to adequately describe the system behavior. (This concept is sometimes referred to as the *law of requisite variety.*) Financial markets are indeed very complex systems, so one needs the ability to develop models with a fairly large number of dimensions (for example, four or more), and therefore a large number of data points are required. Probably there is a real model that if known could predict financial market movement with absolute certainty! This model is often called *God's model,* and He alone knows the true structure of the model. We live in a very complex world, and there are many phenomena that drive financial markets and often in a highly nonlinear fashion. Few people seriously believe it is possible to determine *God's model.*

As a result of the curse of dimensionality, we go into the modeling of financial markets knowing that there is no hope of completely understanding the causes of market behavior. The markets are just too complex for anyone to be able to develop models that give consistently accurate predictions. It is possible, however, to take a statistical perspective of the task. Can predictions be made that have some utility? The predictions have utility if they can be used in such a manner that we gain some edge on random guessing. If the models explain some of the movement in the data, then they might have utility.

In order to develop useful models of very complex systems (like financial markets), we need to propose models with a large number of variables (i.e., dimensions). If there are enough proposed candidate predictors, some of them might turn out to be useful. Unfortunately, there is no guarantee that a useful model can be determined, even if the number of candidate predictors is large (for example, several hundred).

Typically candidate predictors are chosen based on solid economic theory. For example, changes in interest rates from yesterday to today might be expected to have some influence on the change in the price of gold from today to tomorrow. The choice of useful candidate predictors is a subject that continues to receive considerable attention in the literature and is discussed at greater length in Section 2.4.

The preceding paragraph should *not* give the impression that all the candidate predictors will appear in the final model. A typical strategy is to attempt to find a subset of the candidate predictors on which the final model is based. The concept of *data sparseness* was introduced in Section 1.3. As the dimensionality of the model increases by one dimension, the density of the data is halved. As the number of dimensions increases to a value required to develop useful models, often the amount of data becomes too small to support the model! As emphasized in Section 1.3, the model's maximum dimensionality is governed by the data available for modeling. If the amount of data is too small, then strategies to partially compensate for this lack of data are required. Several methods can be used to combat this *curse of dimensionality:*

1. Combine variables in such a manner that one new variable includes effects from several of the original variables.
2. Use a multistage modeling strategy (i.e., the outputs of one stage are the inputs to the next stage).
3. Increase the amount of data by combining similar data sets.

There are well-known techniques for combining variables (e.g., *Principle Components*). Representing high-dimensionality data in fewer dimensions is one of the many interesting topics covered by Dorian Pyle in his book on *Data Preparation for Data Mining.*[12] Multistage modeling is an alternative strategy for accomplishing the same effect. (This subject is discussed in Section 4.6.) Increasing the amount of data can often be accomplished in a straightforward manner. For example, if one is trying to model securities, it might be possible to use groups of similar stocks (e.g., from a particular sector of the stock market). By combining the data from a number of stocks, one can develop a larger database available for modeling. Similarly, modeling of the commodities markets can be based on grouping similar commodities (e.g., currencies).

2.4 CANDIDATE PREDICTORS

One of the first tasks required to model financial markets is to propose a set of *candidate predictors.* If we really knew what drives the market of interest, candidate predictors would not be needed. In reality, no one really knows with certainty what drives financial markets. We might have some idea as to which time series have relevance regarding the market of interest, but that is usually the limit of our knowledge. Knowing that series A, series B, and series C might have some relevance regarding market X is a long way from knowing the connection between A and X, B and X, and C and X. What can be done is to develop a set of candidate predictors based on each of these series and add these predictors to the overall set of predictors that will form the basis of the modeling process.

2.4.1 Differences

The most obvious candidate predictors are simple differences. Assume we are trying to model price changes one time unit into the future. Some obvious choices for candidate predictors are the last changes over N times units (where several values of N might be selected). For example: $N = 1, 2, 3$, and 5. As an example, consider Table 2.1.

In Table 2.1 the Y column is future looking: tomorrow's closing S&P price index minus today's price index. The X columns are all backwards looking. The $X1$ column is the one-day price difference, $X2$ is the two-day price difference, $X3$ is the three-day price difference, and $X4$ is the five-day price difference. Each of the four X columns can be considered as candidate predictors for Y. At the end of each day the X values can be computed. If we could determine a model based on these four X's, then we could predict Y.

To quickly test whether or not these X's by themselves are powerful predictors of Y, we can compute correlations as described in Section 2.1. Based on using 1250 days of data

TABLE 2.1   S&P Prices and Price Differences

DATE	PRICE	Y	X1	X2	X3	X4
970610	865.80	2.95	3.30	19.70	23.70	19.80
970611	868.75	1.60	2.95	6.25	22.65	25.95
970612	870.35	24.35	1.60	4.55	7.85	28.25
970613	894.70	8.65	24.35	25.95	28.90	48.60
970616	903.35	0.65	8.65	33.00	34.60	40.85
970617	904.00	−1.35	0.65	9.30	33.65	38.20
970618	902.65	−3.60	−1.35	−0.70	7.95	33.90
970619	899.05	8.55	−3.60	−4.95	−4.30	28.70
970620	907.60	−3.50	8.55	4.95	3.60	12.90

(920722 to 970630), the results are not very encouraging. The correlation coefficients for Y versus $X1$, $X2$, $X3$, and $X4$ are –0.0548, –0.0554, –0.0726, and –0.0658, respectively. These differences are slightly negatively correlated with Y, but the values are close to zero. (Using a statistical t-test, we can show that these values are not significantly different from zero.)[13] It can be concluded from these numbers that a simple ARMA or ARIMA model will not be sufficient to predict the changes in the S&P price index. However, these X's might combine well with other candidate predictors to create a model that has a greater degree of predictive power.

To test the significance of r, the t statistic is computed: $t = r*sqrt(n - 2)/sqrt(1 - r^2)$ where n is the number of data points used to compute r. If the true value of r is zero (i.e., there is no correlation), then t should be distributed according to the *Student's t* distribution with $n - 2$ degrees of freedom. This distribution rapidly approaches a standard normal distribution: with $n - 2 > 10$, the values are less than 15 percent greater than the standard normal. The largest value of r above (i.e., –0.0726) yields a value of $t = -2.56$. The probability of a value of r being at least this negative (if the true value is zero) is approximately 0.5 percent.

Differences from one series can also be used as candidate predictors for another series. For example, changes in the short-term and long-term interest rates might be used as candidate predictors for an S&P model. Alternatively, one might prefer differences of the ratios of the series. There are many variations to this theme, and only the analyst's imagination limits the number of differences that can be proposed.

2.4.2 Moving Averages

Moving averages by themselves are not useful candidate predictors. If, for example, a time series is trending upward and then suddenly shifts direction, moving averages of the series will also change direction but at a later time. As the time period of the moving average increases, the time lag in this change of direction also increases. This lag in response time makes them quite useless as predictors. However, when used in conjunction with other series, they can be quite powerful.

A typical predictor based on moving averages is the ratio of a series to its moving average. Many trading systems are based on rules concerning such ratios. For example, if the ratio of the closing price of a series to its moving average minus one changes sign, then buy if the change is positive or sell if it is negative. The problem can be stated in mathematical terms: time series such as market prices and interest rates are non-stationary. Moving averages of these time series are also non-stationary, but the ratio of the series to its moving average is a stationary series about a mean value of one. Candidate predictors are useful only if they are based on stationary time series. The need for stationarity of the candidate predictors is due to a basic modeling assumption: past history is relevant to future behavior. If a candidate predictor is nonstationary, then there is a distinct possibility that the current values of the candidate predictor are not within the range of past values. If this is the case, then we can't make a prediction based on similar past values, for none exist!

Other series are similar to moving averages, but they serve slightly different purposes. Two of these series are *moving medians* and *exponential smoothings*. Series based on moving medians take longer to compute than moving averages, but they are less sensitive to outliers in the data. The additional computing time is due to the need to sort the data to compute a median value. Exponential smoothings of data serve a similar purpose as moving averages and moving medians but require less computing time. Instead of N (the number of time periods) used to compute a moving average or a moving median, exponential smoothing uses a smoothing parameter α, which is assigned a value between 0 and 1. We define the exponential smoothing of SERIES as follows:

$$\text{ES\_SERIES}(t) = \text{SERIES}(t) * \alpha + \text{ES\_SERIES}(t - 1) * (1 - \alpha) \quad (2.8)$$

The exponentially smoothed value of the series at time t is computed using only the value of the series at time t and the exponentially smoothed value at time $t - 1$. It can be seen that the contribution of SERIES (t) is the same for both moving averages and exponential smoothing if $\alpha = 1/N$. However, all values of SERIES up to time t affect the exponentially smoothed value. For moving averages and moving medians, only the last N values are used to compute these series.

2.4.3 Moving Slopes

Moving slopes of time series are similar to moving averages. They can be used to identify a trend, but they change direction rather slowly. For example, if a time series is in an upward trend, the value of a moving slope of the series is positive. If the series suddenly changes direction, the moving slope will eventually turn negative but at a later time. The lag in response time increases as the time period used to determine the slope increases. Like moving averages, moving slopes are also nonstationary time series. The value of a moving slope is that it is a measure of the "trendiness" of a time series. It can therefore be used to create stationary candidate predictors that capture this quality.

Once again, the use of moving slopes to create candidate predictors is only limited by the analyst's imagination. A very simple use of a moving slope is to remove the trend from the difference series. Differences are not exactly stationary series, but they are close to being stationary. The series created by subtracting a moving slope from a difference is stationary. Whether it is a useful candidate predictor is another matter.

As an example of a more interesting candidate predictor based on a moving slope, consider two time series: SERIES1 is the N day moving slope of a price series, and SERIES2 is the change in volume of the market in question. The product of these two series might be interesting. This product is not exactly stationary, but since it is based on one-day changes in volume it is fairly close to being stationary. If the volume change is positive and the one-day difference in the price is in the opposite direction to SERIES1, then the market might be changing direction. This candidate predictor is an example of a series that is probably quite meaningless on its own. However, when examined conjointly with other series, it might be useful.

2.5 THE EQUITY CURVE

When modeling financial markets, at some point there is a need to generate equity curves. The purpose of modeling financial markets is to develop trading strategies, and the evaluation of any trading strategy requires analyses of the equity curves. A number of parameters can be computed from an equity curve; these parameters are then used to evaluate the *quality* of the equity curve. Some of the most popular measures of the quality of an equity curve are:

1. The rate of return
2. The Sharpe Ratio
3. The maximum drawdown
4. The average drawdown

The Rate of Return The purpose of a trading system is to generate a return on one's capital. Thus the rate of return (ROI) is probably the most important single measure of performance. Usually, it is stated on an annualized basis. To compute ROI, one looks at the initial and final equity. However, yet another issue must be considered: Is the equity curve based on *compounding?* In other words, if the positions are allowed to grow as the equity grows, then compounding can be assumed. Alternatively, if a standard position size is used, then there is no compounding. Without compounding the computation of ROI is:

$$\text{ROI} = 100*(((\text{equity[final]}/\text{equity[initial]}))*$$
$$\text{RECS\_PER\_YEAR}/\text{RECS})-1) \qquad (2.9)$$

With compounding the computation is:

$$\text{ROI} = 100*((\text{pow}(\text{equity[final]}/\text{equity[initial]}),$$
$$\text{RECS\_PER\_YEAR}/\text{RECS})-1) \qquad (2.10)$$

In these equations, RECS are the number of records in the equity curve. The *pow* function is a standard C function which returns the first argument raised to the power of the second argument. As an example, assume that the equity curve is based on 500 daily records and RECS_PER_YEAR is 250 (i.e., two years of data). Assume that the ratio of equity[final]/equity[initial] is 1.5. Without compounding, ROI is 25 percent; with compounding, ROI is 22.5 percent.

As a more realistic example, consider a portfolio of stocks. If one were to maintain a buy-and-hold strategy, average performance comparable to the S&P index could be expected. Looking at the closing prices for the S&P index from 600104 to 980113, we see that the index rose from 59.91 to 952.12 (i.e., a factor of 15.9 in 38.03 years. On a compounded basis, the ROI for this period is 7.54 percent.

The Sharpe Ratio The Sharpe Ratio as a measure of performance was suggested by William F. Sharpe. His book, *Investments,* summarizes his many contributions to the analysis of investment strategies.[14] Several different definitions of the Sharpe Ratio are in use today, but usually it is defined as the ratio of ROI (expressed as a fraction and not a percentage) to σ (the standard deviation of the fractional equity changes). For this measure of performance to be a dimensionless number, the units for ROI and σ must be the same. To annualize σ, the value of σ generated from the daily fractional equity changes must be multiplied by sqrt(RECS_PER_YEAR). The resulting equation is:

$$\text{Sharpe\_ratio} = (\text{ROI}/100)/(\sigma * \text{sqrt(RECS\_PER\_YEAR)}) \quad (2.11)$$

Based on the S&P data (from 600104 t0 980113), a value of 0.553 is obtained. This is not a very impressive number! A value less than one implies that on the average equity swings are larger than the ROI. After all, the alternative is to invest money in a "risk-free" vehicle (like a CD or a short-term Treasury Bill). When modeling financial markets, the objective is to develop a portfolio that significantly outperforms the risk-free interest rate (which changes over time) but to do it in such a way as to minimize risk. The Sharpe Ratio takes both of these factors into consideration.

The Maximum Drawdown The maximum drawdown is a measure of "pain" as well as performance. The drawdown at any given moment is the fractional decrease in equity from the previous equity high point. As an investor watches an investment lose value, he or she begins to feel pain. The maximum drawdown is a very important measure of this phenomenon. Using a vector language like TIMES [*www.insightware.com*], we can compute the drawdown as follows:

MaxEquity = scan max(Equity)

EquityRatio = Equity/MaxEquity

Drawdown = 1 − EquityRatio

The scan operator creates a series based on the max operator. The first term is Equity[1], the second term is Equity[1] max Equity[2], the third term is Equity[1] max Equity[2] max Equity [3], and so on. The input series is *Equity,* and the resulting series *Drawdown* is a series with the same length as *Equity* and with values from zero to the maximum drawdown. For the S&P series, the maximum observed drawdown was 0.482, which occurred on October 3, 1974. On January 11, 1973, the S&P index fell from a high of over 120 to 62.28 (i.e., a decrease of almost 50 percent). This time period was almost two years in duration and included the Mideast oil crisis.

The Average Drawdown This measure of performance is similar to the maximum drawdown. It is the average value of drawdown over the entire period. For the S&P data, the average drawdown in this period was 0.087. This means that the S&P index over the 38-year period analyzed was on the average down 8.7 percent from its previous high.

A really interesting question is: Can drawdown be predicted? If anyone could predict *when* a serious drawdown was about to occur, he or she would be able to amass a considerable fortune. This is a very difficult prediction problem. However, a relatively simple problem is predicting *the probability of a drawdown of size P percent or greater* at some time in the future. A simple model of financial markets is that the daily fractional changes in equity are normally distributed around a nonzero mean value. If the mean value is positive, then one can expect the equity curve to rise gradually but exhibit periodic upward and downward swings. If the mean value is denoted as μ and the standard deviation as σ, then the equity changes can be said to be distributed as follows:

$$(\text{Equity}[t + 1] - \text{Equity}[t])/\text{Equity}[t] = \mu \pm \sigma \qquad (2.12)$$

Distributions based on this model are often called Inverse Gaussian or Wald distributions.[15] The "motion" derived from this distribution is called *Brownian Motion.* It can be shown[16] that the probability of a drawdown of P percent or greater can be related to μ and σ as follows:

$$\text{Prob(Drawdown} >= P) = (1 - P/100)^{2\mu/\sigma^2} \qquad (2.13)$$

Using the S&P data, the values of μ and σ for the 38.03 years of data are 0.0003263 and 0.008601. The exponent in Equation 2.13 is thus 8.8226. In Table 2.2 the probabilities of drawdowns for various values of P are compared to the actual values obtained from the data.

For this simple Brownian Motion model of the S&P price index, the results are surprisingly accurate. For example, Eq. (2.13) predicts the probability of a 40 percent drawdown as slightly greater than 1 percent. The actual observed number of days with drawdowns of this or greater magnitude was slightly less than 1 percent. Even for a P of 5 percent, the results are still within statistical accuracy. To prove this, a random number generator was used to create a series of the same length (i.e., 9572 records) using the same values of μ and σ. This experiment was repeated 10 times. The average fraction of days with drawdowns greater than or equal to 5 percent was 0.654, with a standard deviation of 0.093. The value of 0.511 is thus well within the 2 sigma range.

As Eq. (2.13) shows, the exponent $2\mu/\sigma^2$ is a powerful parameter that can be used to predict the probabilities of drawdowns of varying magnitude. In Table 2.3 the exponents required to achieve various probability levels are listed for a variety of P values.

As an example of how Table 2.3 is used, consider a requirement that a trading system achieve a probability of less than 0.001 for a 30 percent drawdown. The table shows that an exponent of at least 19.4 is required. If, for example, we could achieve a 20 percent annual ROI, then the daily μ would be 0.00073

TABLE 2.2 Actual and Predicted Fraction of Days with Drawdown >= P (S&P Price Index)

P Values	5%	10%	15%	20%	30%	40%	50%
Actual	0.511	0.347	0.236	0.13	0.026	0.009	0.000
Eq 2.13	0.636	0.395	0.238	0.14	0.043	0.011	0.002

TABLE 2.3   Values of $2\mu/\sigma^2$ Required to Achieve Varying Drawdown Probabilities

P Values	10%	20%	30%	40%	50%
0.0001	87.4	41.3	25.8	18.0	13.30
0.0010	65.6	30.9	19.4	13.5	9.96
0.0100	43.7	20.6	12.9	9.0	6.64

(assuming 250 trading days per year, $0.00073 = 1.2^{-250} - 1$). The target value of σ would thus be sqrt($2\mu/19.4$), which is 0.00867.

2.6 MEASURING THE EFFICIENCY OF A MODELING METHOD

Measurements of efficiency are useful when one is comparing different methods or is doing parameter studies within the domain of a particular method. For example, those using a neural network approach to modeling might want to study the effect of the number of neurons on performance of the network. A very powerful measurement technique is based on the use of artificial data that have been constructed in such a manner as to have the sort of properties that one anticipates in real data sets. If, for example, a modeling method fails using artificial data, it is important to understand the reason for failure before attempting to model real data. Using real data, one can detect failure only if one can be assured that some underlying model drives the data. For financial application this is often not the case. The starting point in generating artificial data sets is first to list the properties one would expect in such data and then to build the artificial data sets accordingly.

To build an artificial data set, one first has to propose a model. Typically, Y (the dependent variable) is created as a function of several X variables (the independent variables) plus random noise. If, for example, the modeling method includes a search through a large candidate predictor space, the data set will include many X variables, most of which are unrelated to Y. The noise component may be generated in a variety of different ways. For example, it might be pure Gaussian random noise (mean of zero and σ of some specified amount); it might be a random value between XMIN and XMAX; or it might be generated by some sort of chaotic process.[17] For financial market modeling, one problem often encountered is a change in market volatility over time. To simulate volatility changes, one might use a Gaussian random noise generator but vary σ over time.

The most straightforward measure of efficiency of the modeling process is the fraction or percentage of the variance of the true signal (i.e., the pure function without noise) that is captured by the process. For example, let's say a data set of 15,000 records has been created in which the Y column is 10 percent signal and 90 percent random noise. Let's assume that 10,000 records are used to create the model and the remaining 5000 are used to test the model. By comparing the actual values of Y with the values of $Ycalc$ (the calculated values of Y for the test set), the VR (Variance Reduction) can be computed using Eq. (1.12). If, for example, a value of VR = 7.25 is computed, then we can say that the modeling process was 72.5 percent efficient for the particular example under consideration. In other words, the modeling process captured 72.5 percent of the actual variance in the pure signal. Note that a perfect model would yield a VR = 10 because the Y column is 10 percent signal and 90 percent noise. Because of the random component in the generated data, it is possible to obtain a value of VR slightly in excess of 10, but if VR significantly exceeds 10 for this example, one would immediately suspect that the process is somehow overfitting the data.

In Section 5.1 a TIMES program for generating artificial data is included in Figure 5.1. This program illustrates the generation of a data set with ten columns of X variables and five Y columns. The X columns are created using a Gaussian random number generator. The first Y column (i.e., column 11) is the pure signal column, and the next four columns are created by adding varying degrees of noise to column 11. Column 12 has a 50 percent noise component, column 13 has 75 percent noise, column 14 has 90 percent noise, and column 15 has 95 percent noise. The pure signal is created using a nonlinear function of $X2$, $X5$ and $X9$. The equations for generating measured levels of noise (i.e., Eqs. [5.1] through [5.4]) are also included in Section 5.1. The results included in Chapter 5 are based on artificial data generated in this manner. In addition, the comparison of KR (kernel regression) and NN (neural networks) included in Appendix D is also based on a similar data set.

NOTES

1. J. Durbin and G. S. Watson, "Testing for Serial Correlation in Least Squares Regression," *Biometrika* 58 (1971): 1–19.

2. J. T. McClave and P. G. Benson, *Statistics for Business and Economics* (New York: Macmillan Publishing, 1994).

3. H. Wold, *A Study in the Analysis of Stationary Time Series* (Stockholm, Sweden: Almqvist & Wiksell, 1938).

4. N. Weiner, *The Extrapolation, Interpolation and Smoothing of Stationary Time Series with Engineering Applications* (New York: John Wiley & Sons, 1949).

5. A. S. Weigend and N. A. Gershenfeld, *Time Series Prediction: Forecasting the Future and Understanding the Past* (Reading, Mass: Addison- Wesley Longman, 1994).

6. M. B. Priestley, *Non-Linear and Non-Stationary Time Series Analysis* (New York: Harcourt Brace & Co., 1988).

7. G. E. P. Box, G. M. Jenkins, G. Reinsel, and G. Jenkins*Time Series Analysis: Forecasting and Control.* (Englewood Cliffs: Prentice Hall, 1994).

8. H. Tong, *Threshold Models in Nonlinear Time Series Analysis* (Springer Verlag, 1983).

9. T. Subba Rao, *Developments in Time Series Analysis* (in honor of M. B. Priestly) (New York: Chapman and Hall, 1993).

10. S. M. Weiss and N. Indurkhya, *Predictive Data Mining: A Practical Guide* (San Francisco, CA: Morgan Kaufman, 1998).

11. R. E. Bellman, *Adaptive Control Processes* (Princeton, NJ: Princeton University Press, 1961).

12. D. Pyle, *Data Preparation for Data Mining* (San Francisco, CA: Morgan Kaufman, 1999).

13. W. Mendenhall and T. Sincich, *Statistics for Engineering and Science,* 3rd ed. (New York: Macmillan, 1992).

14. W. F. Sharpe, *Investments* (Englewood Cliffs, NJ: Prentice Hall, 1979).

15. N. L. Johnson and S. Kotz, *Continuous Univariate Distributions* (Boston: Houghton Mifflin Co., 1970).

16. Private communication, P. Feigin.

17. B. B. Mandelbrot, "A Multifractal Walk Down Wall Street", *Scientific American* (February 1999).

3

KERNEL REGRESSION

3.1 THE BASIC CONCEPT

Kernel regression is one class of data modeling methods that falls within the broader category of *smoothing methods*. The general purpose of smoothing is to find a line or surface that exhibits the general behavior of a dependent variable (let's call it Y) as a function of one or more independent variables. No attempt is made to fit Y exactly at every point. If there is only one independent variable, then the resulting smoothing is a line. If there are more than one independent variables, the smoothing is a surface. Smoothing methods that are based on a mathematical equation to represent the line or surface are called parametric methods. On the other hand, data-driven methods that only smooth the data without trying to find a single mathematical equation are called nonparametric methods. Kernel regression is a nonparametric smoothing method for data modeling.

The distinguishing feature of kernel regression methods is the use of a *kernel* to determine a weight given to each data point when computing the smoothed value at any point on the surface. There are many ways to choose a kernel. Wolfgang Hardle reviews the relevant literature in his book on this subject,[1] and Ullah and Vinod present another overview of the subject.[2]

When using data to create models, it is useful to separate the data into several categories:

45

1. Learning
2. Testing
3. Evaluation (i.e., reserved data for final evaluation)

If the amount of available data is small, then there are strategies for using all the data records to create a model. For such cases, the number of *learning* points is equal to the total number of data points and the number of *testing,* and *evaluation* points is zero. When modeling financial markets, this is rarely the case. There are almost always enough data for both learning and test data sets and usually enough to leave some for final evaluation. The usual strategy is to divide the data, with *nlrn,* *ntst,* and *nevl* points assigned to the three data sets. For various subspaces of the candidate predictor space, the *nlrn* learning points are used to make predictions for the *ntst* testing points, and then some measure of performance is computed. One iterates through spaces following a searching strategy. Only if the final measured performance meets the modeling objectives does one then use the remaining *nevl* points for final *out-of-sample* testing.

To illustrate the procedure for a single one-dimensional space, consider Table 3.1 which includes values for the dependent variable Y and a single independent variable X. In the table we see that *nlrn* is 4 and *ntst* is 3. The four learning points are to be used to smooth the data in such a way as to estimate the Y values for the three test points. Since Y values are already included for the three test points, we can compare the estimated values with the actual values.

Many methods are available for performing this task, but the following discussion is limited to a single variation of kernel regression smoothing. The first decision that must be made is the choice of a kernel. Hardle discusses many alternatives, but for this example a simple exponential kernel is used:

$$w(x_i, x_j, k) = e^{-kD_{ij}^2} \tag{3.1}$$

In this equation the kernel $w(x_i, x_j, k)$ is the weight applied to the ith learning point when estimating the value of Y for the

TABLE 3.1   7 Data Points: 4 in the Learning
Set and 3 in the Test Set

Point	Data Set	X	Y
1	Learning	1	12.0
2	Learning	3	18.0
3	Learning	5	20.0
4	Learning	7	17.0
5	Test	2	14.0
6	Test	4	18.5
7	Test	6	19.0

jth test point. The parameter k is called the *smoothing parameter*, and D_{ij}^2 is the squared distance between the learning and test points. If k is assigned a value of 0, then all points are equally weighted. As k increases, the nearer points are assigned greater weights relative to points further away from the jth test point. As k approaches infinity, the relative weight of the nearest point becomes infinitely greater relative to all other points.

The simplest kernel regression paradigm is what can be called the *Order 0 Algorithm:*

$$y_j = \frac{\sum_{i=1}^{nlrn} w(x_i, x_j, k)\, Y_i}{\sum_{i=1}^{nlrn} w(x_i, x_j, k)} \tag{3.2}$$

In the statistical literature this equation is often referred to as the *Nadaraya–Watson estimator.*[1,2] In this equation y_j is the value of y computed for the jth test point. The values Y_i are the actual values of Y for the learning points. This simple algorithm computes y_j as a weighted average of the Y values of the learning points. As an example, consider only the test point at $x = 2$ (i.e., point 5) in Table 3.2. In the table the kernels required in the calculation are included for $k = 1$ and $k = 0.1$.

From this table and from Eq. (3.2) we can compute two values of $y5$. For $k = 1$ we obtain a value of 15.008, which is very close to the average value of points 1 and 2. The calculation is

TABLE 3.2   Computed Values of $w(x_i, x_5, k)$ for $k = 1$ and $k = 0.1$

Point	X	Y	D_{i5}^2	w for $k = 1$	w for $k = 0.1$
1	1	12	1	0.3679	0.9048
2	3	18	1	0.3679	0.9048
3	5	20	9	0.0001	0.4066
4	7	17	25	1.4e–11	0.0821

dominated by these two points because weights for these points are much greater than the weights for points 3 and 4. For $k = 0.1$, points 3 and 4 have a significant influence on the calculation, and the resulting value is 15.956. The results for all three test points and for both values of k are included in Table 3.3.

The values of y can be used to compare the two alternatives: $k = 1$ and $k = 0.1$. If we choose VR (i.e., Variance Reduction) as the modeling criterion, then Eq. (1.12) is used. In Table 3.3 we use the notation y instead of $Ycalc$ as used in Eq. (1.12). The sums of $(Y(i) - Ycalc(i))^2$ are 1.5002 and 5.3002 for $k = 1$ and $k = 0.1$, respectively. The value of $Yavg$ is 51.5/3 = 17.167, and the sum of $(Y(i) - Yavg)^2$ is therefore 15.167. The values of VR are $100*(1 - 1.5002/15.167) = 90.1$ for $k = 1$ and 65.1 for $k = 0.1$. With so few learning and test points, no significance can be attributed to these results. They do, however, illustrate the basic concept.

It should be emphasized that the values of y (or $Ycalc$) can be computed at any point and not just at the test points. The smoothed curves generated using $k = 1$ and $k = 0.1$ are shown in Figure 3.1. For this example, we see that the curve generated by kernels using $k = 1$ is much closer to the test points then the

TABLE 3.3   Values of y and $(Y - y)^2$ for the Test Points

Point	X	Y	y ($k = 1$)	$(Y - y)^2$	y ($k=0.1$)	$(Y - y)^2$
5	2	14.0	15.008	1.0017	15.956	3.8256
6	4	18.5	18.998	0.2485	17.605	0.8012
7	6	19.0	18.500	0.2500	18.179	0.6734
Sum	Not used	51.5	Not used	1.5002	Not used	5.3002

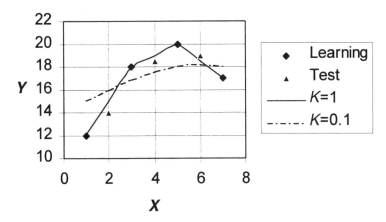

Figure 3.1 Smoothed curves using $k = 1$ and $k = 0.1$.

curve generated using $k = 0.1$. In fact, the smoothed $k = 1$ curve gives the appearance that it actually passes through the learning points. This is not the case. For example, the value at $x = 1$ is 12.11, whereas the actual value of the learning point at $x = 1$ is 12.00. The resolution of the figure is just not enough to clearly show this small difference. The differences for the $k = 0.1$ curve are, however, quite obvious.

Many problems are associated with the use of Eqs. (3.1) and (3.2); these problems are considered in the following sections.

3.2 HIGHER ORDER ALGORITHMS

Equation (3.2) will be referred to as the *Order 0 Algorithm* because it uses a polynomial of Order 0 as the fitting function. (A polynominal of Order 0 is just a constant.) Expanding this notation, we see that the *Order 1 Algorithm* thus uses a polynomial of Order 1. Rather than a weighted average, this algorithm is based on a hyperplane. For a single dimension, the hyperplane is a straight line. In p dimensions it is simply:

$$y = a_1 + a_2\, x_1 + a_3\, x_2 + \ldots + a_{p+1}\, x_p \tag{3.3}$$

Thus for the *Order 1 Algorithm, p* + 1 coefficients are required to define the hyperplane. Clearly, even higher order algorithms can be defined. For example, the *Order 2 Algorithm* is based on a polynomial of Order 2. For a single dimension it is a parabola:

$$y = a_1 + a_2 x_1 + a_3 x_1^2 \qquad (3.4)$$

For two dimensions it includes the interaction term as well as the second-order terms:

$$y = a_1 + a_2 x_1 + a_3 x_2 + a_4 x_1^2 + a_5 x_1 x_2 + a_6 x_2^2 \qquad (3.5)$$

Going to three dimensions, there is the constant term, three first-order terms, three second-order terms, and three interactions terms. There are therefore 10 coefficients (i.e., $a_1, a_2, ..., a_{10}$). As the dimensionality of the space increases, the number of coefficients required to specify the model increases rapidly. The *Order 2 Algorithm* of the space equation in p dimensions is:

$$y = a_1 + \sum_{j=1}^{p} a_{j+1} x_i + \sum_{j=1}^{p} \sum_{k=j}^{p} b_{jk} x_j x_k \qquad (3.6)$$

In general, the number of coefficients required in p-dimensional space for the *Order 2 Algorithm* is $1 + p + p^*(p + 1)/2$. We could continue to even higher order algorithms, but for applications in which there is a substantial noise component in the signal (as in financial market modeling), going to higher order algorithms serves no purpose.

Once an algorithm has been selected, the coefficients are determined for every test point using the learning data points. Typically, the method of linear least squares with weighting is used to determine the N coefficients. We can recast Eqs. (3.3) and (3.6) as follows:

$$y = \sum_{j=1}^{N} A_j g_j(X) \qquad (3.7)$$

Comparing Eqs. (3.3) and (3.7), we see that for the *Order 1 Algorithm*, A_j is just a_j. When comparing Eqs. (3.6) and (3.7), the connection is less obvious for *Order 2*. For example, for two dimensions (i.e., $p = 2$), b_{11} is renumbered as A_4, b_{12} is renumbered as A_5, and b_{22} is renumbered as A_6. In Eq. (3.7), $g_j(X)$ is a function of the vector X. For example, for Eq. (3.3) with $p = 3$, we see that $g_1 = 1$, $g_2 = x_1$, $g_3 = x_2$, and $g_4 = x_3$. For Eq. (3.6) with $p = 3$, the first four g_j's are the same as for Eq. (3.3). The following g_j's are $g_5 = x_1^2$, $g_6 = x_1 x_2$, $g_7 = x_1 x_3$, $g_8 = x_2^2$, $g_9 = x_2 x_3$, and $g_{10} = x_3^2$. Using Eq. (3.7), we find that the least square formulation is the same for both *Order 1* and *Order 2 Algorithms:*

$$CA = V \tag{3.8}$$

A derivation of this equation is included in Appendix A. In this equation C is an N by N matrix and A and V are vectors of length N. The C matrix is symmetrical, and the term C_{jk} is defined as follows:

$$C_{jk} = \sum_{i=1}^{nlrn} w_i \, g_j(X_i) g_k(X_i) \tag{3.9}$$

The term w_i is the weight computed for the ith learning point using, for example, Eq. (3.1). The term X_i is the X vector for the ith learning point. (The reader should note the difference between uppercase X and lowercase x. The lowercase x refers to an element of the X vector.) V vector terms are defined as follows:

$$V_j = \sum_{i=1}^{nlrn} w_i \, g_j(X_i) Y_i \tag{3.10}$$

In this equation Y_i is the value of Y for the ith learning point. Once again, the difference between uppercase and lowercase should be noted. Uppercase Y is used for actual values of Y, and lowercase y is used for calculated values of Y.

To illustrate the process, we can repeat the example from Section 3.1 but using the *Order 1 Algorithm*. Using the data in Table

3.1 and the weights as shown in Table 3.2 for $k = 1$, we must first compute the C matrix and V vector and then the A vector is determined by solving the N linear equations represented by Eq. (3.8). This example is for a one-dimensional space, so N is 2. Using Eq. (3.7), we find that the equation for y is:

$$y = A_1 + A_2 x \qquad (3.11)$$

From this equation we see that $g_1 = 1$ and $g_2 = x$. There is no need to use a subscript for x because we are considering the one-dimensional case. In the following equations x_i refers to the value of x for the ith learning point. The equations for the terms of C and V are therefore as follows:

$$C_{11} = \sum_{i=1}^{4} w_i = 0.7359 \qquad (3.12)$$

$$C_{12} = \sum_{i=1}^{4} w_i x_i = 1.4721 \qquad (3.13)$$

$$C_{22} = \sum_{i=1}^{4} w_i x_i^2 = 3.6819 \qquad (3.14)$$

$$V_1 = \sum_{i=1}^{4} w_i Y_i = 11.0389 \qquad (3.15)$$

$$V_2 = \sum_{i=1}^{4} w_i x_i Y_i = 24.2924 \qquad (3.16)$$

The two equations to be solved are:

$$\begin{aligned} C_{11} A_1 + C_{12} A_2 &= V_1 \\ C_{21} A_1 + C_{22} A_2 &= V_2 \end{aligned} \qquad (3.17)$$

Since the C matrix is symmetric, $C_{21} = C_{12}$. Solving Eq. (3.17), the resulting values of A_1 and A_2 are 9.0034 and 2.9980. Using

TABLE 3.4 *Order 1* Results for Three Test Points

Point	X	Y	A_1	A_2	Ycalc
5	2	14.0	9.0034	2.9980	14.999
6	4	18.5	15.0005	0.9995	18.999
7	6	19.0	27.4841	−1.4975	18.499

Eq. (3.11) with $X=2$, we get a value of y equal to 14.999 (i.e., 9.0034 + 2*2.9980). The results for all three test points are summarized in Table 3.4.

The point to note in examining Table 3.4 is that a separate line (i.e., separate values of A_1 and A_2) is computed for each test point. The differences from point to point are due to the different weights assigned to the learning points based on the distances from the test points (see Eq. 3.1). However, if the value of k in Eq. (3.1) is 0, then all points are equally weighted and the values of the A_k's are constant for all test points. This point is illustrated using the data in Table 3.5.

Assuming all learning points are given the same weight (e.g., w_i = 1), results for all three algorithms are included in Table 3.6.

TABLE 3.5 Eleven Data Points: Eight in the Learning Set and Three in the Test Set

Point	Data Set	x1	x2	Y
1	Learning	1	−6	5
2	Learning	3	8	−11
3	Learning	5	0	9
4	Learning	7	2	−3
5	Learning	9	−5	53
6	Learning	11	7	−57
7	Learning	13	3	−19
8	Learning	15	−1	31
9	Test	2	0	5
10	Test	6	4	−11
11	Test	10	−2	31

TABLE 3.6   Results for Test Points Using Three Algorithms with Equal Weighting

Point	x1	x2	Y	y0	y1	y2
9	0	5	5	1.000	1.715	5.249
10	6	4	−11	1.000	−15.450	−11.167
11	−2	3	31	1.000	17.450	31.407

In this table the $y0$ values are the values computed using the *Order 0 Algorithm,* and $y1$ and $y2$ are the results using the *Orders 1 and 2 Algorithms.*

The results in Table 3.6 illustrate several points. Notice that all three values of $y0$ are the same (i.e., exactly 1). The value 1 is just the average value of Y for the eight learning points included in Table 3.5. In other words, if all learning points are equally weighted, *Order 0* yields a single value for all test points. At first glance this appears to be a useless result. However, as we will see in Section 3.3, unit weighting can prove to be quite advantageous, even if the *Order 0 Algorithm* is used.

The results in Table 3.6 for the *Order 1 and 2 Algorithms* (i.e., $y1$ and $y2$) do vary from test point to test point, even though all learning points were equally weighted. The two equations obtained were:

$$y1 = 0.2848 + 0.7152\, x_1 + 5.0065\, x_2 \qquad (3.18)$$

$$y2 = 3.0488 + 1.1200x_1 + 0.8039x_2 - 0.0100x_1^2 - 0.9889x_1x_2 - 0.0037x_2^2 \qquad (3.19)$$

Note that both of these equations apply to all three test points. Comparing the values of $y0$, $y1$, and $y2$ to Y in Table 3.5, we see that the values of $y2$ are much closer than either the values of $y0$ and $y1$. One should not conclude that *Order 2* is always preferable to *Order 0* or *Order 1*. For this particular example, the Order 2 polynomial turned out to yield the best results.

3.3 THE BANDWIDTH CONCEPT

If we consider development of a one-dimensional model (i.e., y as a function of x), and if we use a kernel-smoothing algorithm then the *bandwidth* concept refers to the definition of the kernel (e.g., Equation 3.1). The bandwidth h is defined as a region around a test point in which only learning points that fall within this region are given nonzero weights. For a test point located at x_j only learning points that fall within the region $x_j - h/2$ to $x_j + h/2$ are used for the estimation of y. If no learning points fall within this region, then y cannot be estimated. As a result of this problem, another approach called the *K-nearest neighbor estimate (K − N/N)* is often used.[1] For smoothing problems in which the learning values of x may be chosen, a constant bandwidth is perfectly acceptable. Even if the x values cannot be chosen, if there is high data density, then a constant bandwidth is still a reasonable choice. If $K − N/N$ is used, it can be considered a variable bandwidth approach. In other words, for every test point, the bandwidth is chosen so that the K closest learning points are given nonzero weights.

Can we extend the bandwidth concept to higher dimensions? Equation (3.1) provides the clue. In this equation the kernel is based on a squared distance. We can easily define a distance in p-dimensional space and base bandwidth on this distance. It should be clear that if the x variables in p-dimensional space have radically different scales, then a straight Cartesian distance is useless. The long dimension would totally dominate the calculation. Typically, some sort of normalized distance is used when considering kernel regression in p-dimensional space (where p is greater than one). The most common methods of normalizing distances are to divide the actual values for each dimension by either the range or standard deviation of the dimension. In p-dimensional space the value of D_{ij} (i.e., the normalized distance between the jth test point and the ith learning point) is computed as follows:

$$D_{ij}^2 = \sum_{d=1}^{p} \left((x_{di} - x_{dj})\big/ N_d\right)^2 \tag{3.20}$$

The value of D_{ij} is just the square root of D_{ij}^2. However, it is usually more advantageous to work with D_{ij}^2 because it eliminates the need for taking square roots. If the numbers of learning and test point are large, then avoiding the square root calculation can save significant compute time. In this equation x_{di} is the dth dimension of the ith learning point and x_{dj} is similarly defined for the jth test point. The N_d term is the normalization constant for the dth dimension (e.g., the range for dimension d).

The primary reason for introducing the bandwidth concept is to eliminate the effect of very distant points on the estimate at a given test point. This can be accomplished by using a large smoothing constant. For example, using Eq. (3.1), we see that the choice of a large value of k will also eliminate the influence of distant points. However, this might be counterproductive because a large value of k also reduces the smoothing, and in the extreme, the estimates are based primarily on the single nearest neighbor of each test point.

Using bandwidth (or $K - N/N$) also saves a considerable amount of compute time. Defining the number of learning and test points as N_l and N_t, if all learning points are used to determine all test points, the computational complexity is $O(N_l * N_t)$. In other words, the time to complete the computation increases as the product of N_l and N_t. For modeling of financial markets, these numbers can be very large and therefore some approach to reducing computational complexity is necessary. We will see in Chapter 4 that high-performance kernel regression systems are based on a variation of the bandwidth concept.

3.4 ERROR ESTIMATES

In Sections 3.1 and 3.2 *Order 0, 1,* and *2 Algorithms* were described. The errors associated with the predicted values of Y can easily be computed. Typically, predicted errors are expressed as σ_y, which is the estimated standard deviation of the prediction. The predicted value of σ_y for *Order 0* is simply the standard deviation of the weighted average:

$$\sigma_y^2 = \frac{\displaystyle\sum_{i=1}^{n} w_i \left(Y_i - y_j \right)^2}{(n-1)\displaystyle\sum_{i=1}^{n} w_i} \qquad (3.21)$$

In this equation y_j is the predicted value of the jth test point and is computed as the weighted average of the n values of Y used to make the prediction (i.e., Eq. 3.2). For every test point there is a value of σ_y. The equation is valid with or without a bandwidth limitation on the choice of data points. Using a bandwidth (or some similar device to reduce the number of data points) reduces the value of n. Notice that as n decreases, the term $(n-1)$ in the denominator decreases and tends to increase σ_y. However, as n decreases, distant points are rejected and one would expect that the nearer values of Y_i would be closer to the actual value of Y at the test point. Thus, one would expect that there is an optimum value of n (or bandwidth) at which the value of σ_y is minimized.

Derivation of Eq. (3.21) follows naturally from the more general derivation of σ_y for all three algorithms. The three algorithms are based on three different fitting functions. The fitting function for *Order 0* is a constant:

$$y = a_1 \qquad (3.22)$$

Order 1 uses a hyperplane in p-dimensional space (Eq. 3.3):

$$y = a_1 + a_2 x_1 + a_3 x_2 + \ldots + a_p + x_p \qquad (3.3)$$

Order 2 uses a complete second-order multinomial (Eq. 3.6):

$$y = a_1 + \sum_{j=1}^{p} a_{j+1} x_i + \sum_{j=1}^{p}\sum_{k=j}^{p} b_{jk} x_j x_k \qquad (3.6)$$

All three equations can be recast into the form of Eq. (3.7).

$$y = \sum_{j=1}^{N} A_j g_j (X) \qquad (3.7)$$

For *Order 0*, $N = 1$ regardless of the dimensionality of the space. For *Order 1*, $N = p + 1$ for a p-dimensional space. For *Order 2*, $N = 1 + p + p^*(p + 1)/2$. The functions $g_j(X)$ can be determined by comparing Eqs. (3.2), (3.3), and (3.6) with Eq. (3.7). Examples for *Order 1* and *2* are included in Section 3.2. For *Order 0* the function g_1 is 1.

In Section 3.2 the values of the A_j's were determined by solving the matrix Eq. (3.8). The elements of the C matrix were determined using Eq. (3.9). Using this same formulation, Eq. (3.2) (the order 0 prediction for test point j) can be derived. Since $N = 1$, Eq. (3.8) is a single equation:

$$C_{11} A_1 = V_1 \tag{3.23}$$

where

$$C_{11} = \sum_{i=1}^{n} w_i g_1 g_1 = \sum_{i=1}^{n} w_i \tag{3.24}$$

$$V_1 = \sum_{i=1}^{n} w_i g_1 Y_i = \sum_{i=1}^{n} w_i Y_i \tag{3.25}$$

Substituting (3.24) and (3.25) into (3.23), we solve for A_i. The resulting equation for A_i is the same as Eq. (3.2) and is the predicted value of Y (according to Eq. 3.22). Thus the general formulation used in Section 3.2 for *Orders 1* and *2* is also applicable to *Order 0*.

The method of least squares includes a general formulation for σ_y:[3,4]

$$\sigma_y^2 = \frac{S}{n - N} \sum_{j=1}^{N} \sum_{k=1}^{N} g_j(X) g_k(X) C_{jk}^{-1} \tag{3.26}$$

The value of σ_y is thus determined by taking the square root of Eq. (3.26). In this equation, C^{-1} refers to the inverse matrix of C. The term C^{-1}_{jk} is therefore the term on the jth row of the kth column of the inverse matrix. The symbol S refers to the sum of the residuals and is computed as follows:

$$S = \sum_{i=1}^{n} w_i (Y_i - y_i)^2 \qquad (3.27)$$

The values of Y_i are the actual values of Y for the learning points, and the values of y_i are the computed values of Y for the learning points (using either Eq. 3.22, 3.3, or 3.6). We can show that Eq. (3.26) is equivalent to Eq. (3.21) for *Order 0* by substituting (3.27) into (3.26) and noting that $N = 1$ and $g_1(x) = 1$. Since $N = 1$, C_{11}^{-1} is a scalar and is just $1/C_{11}$.

For *Orders 1* and *2*, the C matrix is computed in order to determine the values of y_j. The C matrix can be inverted to solve Eq. (3.8), however, this is not the fastest way to solve simultaneous linear equations.

An additional cost is required if this matrix is inverted; therefore, in high-performance systems, one would generally invert the matrix only for final results in which the values of σ_y are required. For *Order 1* with $p = 1$ (i.e., one dimensional), $N = 2$. For this case, Eq. (3.26) reduces to the following:

$$\sigma_y^2 = \frac{S}{n-2}(C_{11}^{-1} + 2x_j C_{12}^{-1} + x_j^2 C_{22}^{-1}) \qquad (3.28)$$

The constant 2 associated with the $C^{-1}{}_{12}$ term is due to the fact that the C^{-1} matrix is symmetric, and therefore $2C^{-1}{}_{12}$ is used instead of $(C^{-1}{}_{12} + C^{-1}{}_{21})$. From this equation we see that the value of σ_y is different at every x_j (i.e., the x value of the jth test point). For *Order 2* with $p = 1$ (i.e., one dimensional), $N = 3$. For this case, Eq. (3.26) reduces to the following:

$$\sigma_y^2 = \frac{S}{n-3}(C_{11}^{-1} + 2x_j C_{12}^{-1} + 2x_j^2 C_{13}^{-1} + x_j^2 C_{22}^{-1} + 2x_j^3 C_{23}^{-1} + x_j^4 C_{33}^{-1}) \qquad (3.29)$$

For *Order 1* with $p = 2$ (i.e., two dimensional), $N = 3$. For this case, Eq. (3.26) reduces to the following:

$$\sigma_y^2 = \frac{S}{n-3}\left(\begin{array}{l} C_{11}^{-1} + 2x_{1j} C_{12}^{-1} + 2x_{2j} C_{13}^{-1} + x_{1j}^2 C_{22}^{-1} + \\ 2x_{1j} x_{2j} C_{23}^{-1} + x_{2j}^2 C_{33}^{-1} \end{array} \right) \qquad (3.30)$$

As p increases, Eq. (3.26) becomes increasingly cumbersome, and there is no justification for expressing the equation explicitly. In any computer code in which this equation is used, a modified version of the loop form as expressed in (3.26) is preferable to explicit forms such as (3.28) to (3.30). The equation can be modified to eliminate the need to compute terms below the diagonal of the matrix.

To illustrate this calculation, consider the data included in Table 3.5. This data includes eight learning points and three test points. The data is two dimensional (i.e., $p = 2$). The values of y computed using the three algorithms are included in Table 3.6. At this point we will compute the values of σ_y for the test points for each of the three algorithms. To simplify the calculation, we will again assume that all learning points are equally weighted (i.e., $w_i = 1$). The three fitting functions for this problem are:

Order 0: $\qquad\qquad\qquad y = A_1$ $\qquad\qquad\qquad$ (3.31)

Order 1: $\qquad y = A_1 + A_2 x_1 + A_3 x_2$ $\qquad\qquad$ (3.32)

Order2: $\;\; y = A_1 + A_2 x_1 + A_3 x_2 + A_4 x_1^2 + A_5 x_1 x_2 + A_6 x_2^2$ $\quad$ (3.33)

For *Order 0,* the C matrix contains only one element and is computed using Eq. (3.24):

$$C_{11} = \sum_{i=1}^{8} w_i = 8 \qquad\qquad (3.34)$$

The term C^{-1}_{11} is thus 0.125. The value of S is computed using Eq. (3.27), and a result of 7608 (i.e., $(5-1)^2 + (-11-1)^2 + (9-1)^2 \ldots$) is obtained. The value of σ_y for all three test points is determined using Eq. (3.26). Noting that $N = 1$ and $g_1 = 1$, we obtain the following equation:

$$\sigma_y^2 = \frac{S}{n-1} C_{11}^{-1} = \frac{7608}{7} * 0.125 = 135.86 \qquad (3.35)$$

The computed value of σ_y is thus 11.65 and is the same for all three test points (since the value of y_j is the same [i.e., 1] for all three points).

For *Order 1,* the C matrix is 3 by 3 and therefore contains nine elements. The elements are computed according to Eq. (3.9). Using $w_i = 1, g_1 = 1, g_2 = x_1$, and $g_3 = x_2$, we obtain the following matrix:

$$C = \begin{Bmatrix} 8 & 64 & 8 \\ 64 & 680 & 88 \\ 8 & 88 & 188 \end{Bmatrix} \qquad (3.36)$$

Inverting this matrix, we obtain the C^{-1} matrix:

$$C^{-1} = \begin{Bmatrix} 0.5061 & -0.04773 & 0.0008091 \\ -0.04773 & 0.006068 & -0.0008091 \\ 0.0008091 & -0.0008091 & 0.005663 \end{Bmatrix} \qquad (3.37)$$

Since all learning points are equally weighted, one hyperplane is determined and is applicable to all three test points. The coefficients of the plane are included in the equation for $y1$ in Eq. (3.18) (i.e., 0.2848, 0.7152, and 5.0065). Using this equation, we determine S according to Eq. (3.27).

$$S = \sum_{i=1}^{n} w_i [Y_i - (A_1 + A_2 x_1 + A_3 x_2)]^2 \qquad (3.38)$$

The value obtained for S is 3182.3. The values of σ_y can now be computed from Eq. (3.30). For example, for Point 9 in Table 3.5, the value of x_1 is 2 and the value of x_2 is 0. The resulting equation is:

$$\sigma_y^2 = \frac{S}{n-3} (C_{11}^{-1} + 4 C_{12}^{-1} + 4 C_{22}^{-1}) = \frac{3182.3 * 0.3395}{5} = 216.0 \qquad (3.39)$$

The value of σ_y is thus 14.7. The values of σ_y for Points 10 and 11 are determined in a similar manner, and both values are 11.56. For both of these points, the equations are more complicated than Eq. (3.39) because the terms containing x_2 must be included.

TABLE 3.7   Values of σ_y for Three Algorithms Using Data from Table 3.5

Point	x1	x2	Y	σ_y (Order 0)	σ_y (Order 1)	σ_y (Order 2)
9	2	0	5	11.65	14.70	1.49
10	6	4	−11	11.65	11.56	0.90
11	10	−2	31	11.65	11.56	0.91

For *Order 2,* the C matrix is 6 by 6. In a similar manner, we obtain the following matrix:

$$C = \begin{Bmatrix} 8 & 64 & 8 & 680 & 88 & 188 \\ 64 & 680 & 88 & 8128 & 888 & 1152 \\ 8 & 88 & 188 & 888 & 1152 & 548 \\ 680 & 8128 & 888 & 103496 & 9784 & 10508 \\ 88 & 888 & 1152 & 9784 & 10508 & 4360 \\ 188 & 1152 & 548 & 10580 & 4360 & 8516 \end{Bmatrix} \qquad (3.40)$$

The calculation proceeds in a manner similar to the *Order 1* calculation. In place of Eq. (3.32) in (3.38), Eq. (3.33) is used. The resulting value of S is 4.044. The resulting values of σ_y for the three test points are 1.49, 0.90, and 0.91. These values are considerably less than the values obtained using the *Orders 0* and *1 Algorithms.* This is not a surprising result if one examines Table 3.6 and notices how much closer the values of $y2$ are to the Y values as compared to $y0$. The results are summarized in Table 3.7.

3.5 APPLYING KERNEL REGRESSION TO TIME SERIES DATA

In the previous sections, kernel regression was applied to data in which there was no special significance to the order in which the records were analyzed. In most financial applications, however, the order of the data is extremely important. For example, daily data is ordered by date, and intraday data is ordered by date/time. There are other possible orderings of the data. For example, if one wishes to analyze a group of stocks using daily data, one might order the data on the basis of date/stock_code. Once we acknowledge that the order of the data is important, then the choice of learning and test data sets becomes crucial.

The time dimension introduces another level of complexity to the analysis: *how much importance do we attach to recent data records as opposed to earlier records?* Is there a simple way to take this effect into consideration? Common sense leads us to the basic conclusion that if we are to predict a value of Y at a given time, we should only use learning data from an earlier time.[5] Using this principle, one would choose learning data as the earlier records and then use the later records for testing. But this procedure tends to be overly restrictive. For example, if we had 10 years of daily data from 1989 through 1998 (approximately 2500 records of data), and if we wanted to test on about one-third of the records, then the last three or four years would be used for testing. But this seems to be unrealistic, particularly for the later test data records. We would, for example, be modeling records in 1998 using data from 1989 to about 1994 and not including the more recent records in the modeling process.

This problem has a simple solution: All that one must do is to make the learning data set dynamic. In other words, once a record has been tested, it is then available for updating the learning data set prior to testing the next record. The analyst can allow the learning data set to grow, or, alternatively, for each record added, the earliest remaining record in the learning set can be discarded. These two alternatives will be referred to as the *growing* option and the *moving window* option. For cases in which the learning set is not changed, we will refer to this alternative as the *static* option.

To illustrate these three alternatives, consider the data in Table 3.5 and the results included in Table 3.6. The results in Table 3.6 were computed using the *static* option. In other words, learning Points 1 through 8 were used to compute predicted values of Y for test Points 9, 10, and 11. The three values included in Table 3.6 for each test point are the predicted value using the three algorithms (i.e., *Orders 0, 1,* and *2*). We can repeat the calculations using the *growing* and *moving window* options.

The results for the *growing* option are included in Table 3.8.

TABLE 3.8   Results for Test Points Using the *Growing* Option

Point	x1	x2	Y	y0	y1	y2
9	0	5	5	1.000	1.715	5.249
10	6	4	−11	1.444	−14.974	−11.214
11	−2	3	31	0.200	17.695	31.435

We see that the results for Point 9 are exactly the same as those for the *static* option (i.e., Table 3.6). However, Point 9 is then added to the learning set for the calculations for Point 10, and both Points 9 and 10 are used in the calculations for Point 11. The values of $y0$ can most easily be verified. Since all points are weighted equally, the value of $y0$ for Point 10 is just the average value of Y for Points 1 through 9 (i.e., [8 * 1.000 + 5]/9 = 1.444). The value for Point 11 is the average value for Points 1 through 10 and is computed in a similar manner (i.e., [8* 1.000 + 5 − 11]/10 = 0.200). Using the *static* option with equally weighted learning points, we determined a single plane and used it to compute all three values of $y1$ in Table 3.6. One single plane is no longer applicable when the *growing* option is used. The coefficients of the plane must be recomputed for each test point. Similarly, the coefficient used in the computations of the values of $y2$ must also be recomputed for each test point. The results for the moving option are included in Table 3.9.

We first note that the results for Point 9 are exactly the same as those observed in both Tables 3.6 and 3.8: a perfectly reasonable outcome. At first glance, the results for $y0$ for Points 10 and 11 seem strange: they are exactly the same as the results in Table 3.6 (i.e., using the *static* option). However, inspection of Table 3.5 provides the explanation. The computa-

TABLE 3.9   Results for Test Points Using the *Moving Window* Option

Point	x1	x2	Y	y0	y1	y2
9	0	5	5	1.000	1.715	5.249
10	6	4	−11	1.000	−12.664	−11.379
11	−2	3	31	1.000	30.506	31.720

tion for Point 10 requires discarding Point 1 and adding Point 9 to the learning data set. Since the Y values for both of these points are the same (i.e., 5), the average value of Y remains the same. Since $y0$ is just the average value of Y, it remains 1. Similarly, Point 11 is computed by discarding Points 1 and 2 and adding Points 9 and 10. Since the Y values of Points 2 and 11 are the same (i.e., -11), we once again get an average value of 1. However, the values of $y1$ and $y2$ for Points 10 and 11 are different from both the *static* and *growing* option results. The computation of $y1$ and $y2$ requires not only the Y values of the learning data points but also the values of $x1$ and $x2$. Since a different set of values is used for each test point, the coefficients change from test point to test point and the results are different from the results obtained using the other options.

3.6 SEARCHING FOR A MODEL

Typically, when modeling financial data, the analyst proposes a large number of potentially useful candidate predictors. Depending on the problem and the available computer resources, the number of candidate predictors can range into the hundreds. The strategy proposed for such analyses is first to try to build a model using each candidate predictor individually. Once all such 1D (i.e., one-dimensional) spaces have been considered, the analysis proceeds to 2D spaces, then 3D spaces, and so on. If one considers all possible combinations, the number of spaces to be examined explodes exponentially as the number of candidate predictors increases.

To illustrate the magnitude of the problem, we need to be able to compute C_d^n, which is the number of combinations of n things taken d at a time. For our purposes, n is the number of candidate predictors and d is the dimensionality of the spaces. All books on probability theory include the following equation for C_d^n:

$$C_d^n = \frac{n!}{d!(n-d)!} \tag{3.41}$$

For example, if $n = 100$ and $d = 2$, the number of possible 2D spaces is $100!/(2! * 98!)$ which reduces to $100*99/2 = 4950$. If we plan to examine all possible spaces starting from 1D spaces up to a dimensionality of $dmax$, then S_{total} (the total number of spaces is simply:

$$S_{total} = \sum_{d=1}^{dmax} C_d^n \qquad (3.42)$$

Table 3.10 includes values of S_{total} for various combinations of n and $dmax$. It is clear from this table that exhaustive searches for all combinations of candidate predictors become increasingly costly as n and $dmax$ increase. What is required is a searching strategy that limits the number of combinations to be examined to a reasonable number. The definition of *reasonable* is, of course, both problem and machine dependent. One can quickly estimate the approximate time to analyze a single space, and knowing how much time we wish to devote to the analysis, we can compute the number of spaces that can be examined in the desired time.

The simplest searching strategy is to use a *forward stepwise* approach. One would simply locate the best candidate predictor using a 1D analysis and then pair this *best predictor* with all others to find the *best pair*. Using the *best pair*, one would then proceed to examine all triples that can be made from the *best pair* to obtain the *best triple*. This procedure can be carried on to higher and higher dimensions. The total amount of spaces examined using this procedure is very small. For example, using this very simple *forward stepwise* approach with $n = 200$ and $dmax = 5$, we would only have to examine $200 + 199 + 198 + 197 + 196 = 990$ spaces. From Table 3.10 we see that an exhaustive search of all combinations requires examining over 2.6 billion spaces! What is really required is a strategy that is a compromise between these two extremes.

A simple modification of the *forward stepwise* approach is to allow a user-specified number of spaces to survive each level of dimensionality. These *survivors* will then be used with all other candidate predictors to create spaces at the next higher dimen-

TABLE 3.10 Values of S_{total} for Combinations of n and $dmax$

n	dmax = 1	dmax = 2	dmax = 3	dmax = 4	dmax = 5
5	5	15	25	30	31
10	10	55	175	385	637
20	20	210	1350	6195	21699
50	50	1275	20875	251175	2369935
100	100	5050	166750	4087975	79375495
150	150	11325	562625	20822900	612422930
200	200	20100	1333500	66018450	2601668490

sion. For example, let's assume a parameter called *num_survivors(d)*, which is the number of survivors that will be used to create spaces at dimension $d + 1$. Assume $n = 200$ and *num_survivors(1)* is specified as *10*. After the 200 1D spaces are examined and the 10 best performers are noted, the number of 2D spaces to be examined is 1945. This number includes all combinations of the 10 best (i.e., 10*9/2 = 45) plus all combinations of each of the 10 best with all the remaining 190 (i.e., 10*190 = 1900).

Calculation of the number of 3D spaces that must be examined is complicated by the fact that some predictors might appear in more than one of the best 2D spaces. For example, let's continue the previous example using $n = 200$ and set *num_survivors(2)* to 3. Let's also assume that the best three pairs are ($X7$, $X19$), ($X21$, $X47$), and ($X36$, $X52$). The number of 3D spaces that would have to be examined would be 3*198 = 594. However, if the third space was ($X47$, $X52$), then the 3D space ($X21$, $X47$, $X52$) would be examined twice unless the search algorithm included some mechanism for preventing duplicate examinations. In other words, there are only 593 different 3D spaces for this case. We can, however, state an upper limit upon $S(d + 1)$, the number of spaces that will be examined at dimensionality $d + 1$ using this algorithm:

$$S(d + 1) \leq num\_survivors(d) * (n - d) \qquad (3.43)$$

This equation gives the analyst a quick method for estimating the total number of spaces that will be examined. For example, consider the case of $n = 200$ and $dmax = 5$. Assume that *num_survivors(d)* is set to 10 for all values of d. There are 200 1D spaces, and we have already computed $S(2) = 1945$. We can use Eq. (3.43) to estimate upper limits for $S(3)$, $S(4)$, and $S(5)$: $S(3) < = 10*198 = 1980$, $S(4) < = 10*197 = 1970$, and $S(5) < = 10*196 = 1960$. The upper limit for S_{total} for this example is 8055. This number of spaces is about eight times greater than the number of spaces examined using a simple *forward stepwise* search (i.e., 990), but it is still many orders of magnitude less than the number of spaces that would be examined if all possible combinations were considered.

Any analysis must include a definition of the *modeling criterion MC*. A number of possible definitions of *MC* were discussed in Section 1.4, but this list is by no means exhaustive. For example, when modeling financial markets, one might prefer some sort of criterion based on trading performance. However, regardless of the choice of *MC*, we end up with a single value for each space and spaces can be graded on the basis of this value. The best space is simply the space with the highest value of *MC*. A search algorithm should include some sort of criterion for aborting the search when it becomes pointless to proceed. The choice of a maximum value of *dmax* is really quite arbitrary. What we would like to choose is a dimensionality that is high enough to capture the really good model (or models) if such models really exist but on the other hand not so high that the data density is absurdly low (see Section 1.3).

One approach to this problem is to treat the parameter *num_survivors(d)* as an upper limit. An added criterion for evaluating a space might be the required improvement from one dimension to the next higher dimension. If we define this required improvement in *MC* as δ, then a space failing to meet this criterion is immediately rejected regardless of the measured value of *MC*. For example, assume that one of the survivors of the 2D analyses is the space $(X7, X19)$ and the measured value of *MC* for this space is 5.78. Assume that $\delta = 1$ and the 3D spaces $(X7, X11, X19)$ and $(X7, X19, X46)$ are the best two 3D spaces created from $(X7, X19)$. Furthermore, the values of *MC* for these two spaces are 7.31 and 6.49, respectively. The first of these two spaces would be included in the list of possible survivors, but the second would be immediately rejected. The first space (i.e., $(X7, X11, X19)$) would only become a survivor if the value 7.31 turned out to be within the top *num_survivors(3)* of 3D spaces examined. If there are no survivors for a particular level of dimensionality d, then the search is aborted even if $d<dmax$.

3.7 TIMING CONSIDERATIONS

Use of kernel regression in data modeling for the types of problems associated with financial markets requires careful consideration of computational time. When one is faced with the task of developing prediction models in which there are thousands of data records and hundreds of candidate predictors, computational efficiency is of utmost importance. For computers exploiting a single processor, the total time for an analysis T_{total} can be estimated as follows:

$$T_{total} = S_{total} * T_{avg} \tag{3.44}$$

In this equation S_{total} is the total number of spaces examined and T_{avg} is the average time required per space. (In Section 4.9, parallel processing is considered, and the implications regarding this equation are discussed.) In the previous section, some aspects related to controlling S_{total} were considered. In this section the emphasis is on controlling T_{avg}.

In Section 3.1, the basic concept of a kernel regression analysis was described. Values of the dependent variable Y were predicted for *ntst* test points using *nlrn* learning points. Using this very simple approach to kernel regression, we see that the time required for evaluating a single space would be $O(nlrn*ntst)$. In other words, the value of T_{avg} would tend to be proportional to the product *nlrn*ntst*. If we assume that both *nlrn* and *ntst* are proportional to *ntot* (i.e., the total number of data records available for the analysis), we see that T_{avg} would thus be proportional to $ntot^2$. For typical cases in which the value of *ntot* is several tens of thousand (or even hundreds of thousands if intraday data is used), the values of T_{avg} become intolerably large.

In Section 3.3 we discussed the bandwidth concept in which only nearby learning points are used to predict values of Y for each test point. A simple way of accomplishing this is to compute the distance from every learning point to each test point and use only those points within a specified distance. However, the computation of all these distances is still $O(nlrn*ntst)$. It is true that time will be saved because the remainder of the calculation will be faster. However, we are still left with a term that is increasing as $ntot^2$, and eventually this term will dominate the time required per space.

Yet another major problem is created by choosing a maximum distance between learning and test points. For most problems, the data density is not even approximately constant throughout a particular space. In other words, in some regions within the space many learning points are concentrated, and in other regions they are sparsely populated. If we must select one distance for the entire space, then it will yield estimates of Y for some test points based on many learning points and some estimates based on few points. It is also possible that some test data points will fall within regions where none of the learning points is within the specified distance.

A simple solution to this problem is to specify a new parameter *numnn,* which is the *number of nearest neighbors* that must be used for each test point. To accomplish this in the simplest and most straightforward manner, for each test point one would first compute all the distances to the learning points and then sort the distances. The closest *numnn* learning points would then be used to compute the estimated value of Y. The time required to sort *nlrn* distances is $O(nlrn*log(nlrn))$. Since we would require *ntst* such sorts, we would be left with a term that is $O(ntst*nlrn*log(nlrn))$. This simple solution is thus plagued with a very high computational cost. What is required is some method of controlling the choice of learning points for each test point in such a manner that we achieve this in a rapid and efficient manner. Alternative approaches to this problem are considered in Chapter 4.

NOTES

1. W. Hardle, *Applied Nonparametric Regression* (Cambridge, UK: Cambridge University Press, 1990).

2. A. Ullah and H. D. Vinod, "General Nonparametric Regression Estimation and Testing in Econometrics," *Handbook of Statistics 11* (North Holland, 1993).

3. J. Wolberg, *Prediction Analysis* (New York: Van Nostrand Reinhold, 1967).

4. P. Gans, *Data Fitting in the Chemical Sciences* (New York: John Wiley & Sons, 1992).

5. When the amount of data is relatively small or if there is reason to believe that the model is relatively invariant with time, the analyst might choose to use future data to make predictions on the past.

4

HIGH-PERFORMANCE
KERNEL REGRESSION

4.1 SOFTWARE CONSIDERATIONS

The previous chapter described the fundamental concepts related to the kernel regression method of data modeling, and Section 3.7 explained the need for efficiency on the basis of computational complexity. Because of the very nature of modeling financial markets (i.e., thousands of data records and perhaps hundreds of candidate predictors), the need for computational speed is paramount. The entire argument for speed is summarized by Eq. (3.44): if S_{total} (the number of spaces that must be examined) is large, then T_{avg} (the average time for examining a single space) must be small enough so that the total time required for a complete analysis is reasonable.

The bandwidth concept described in Section 3.3 hints at the type of solution that should yield impressive improvements in speed. If we can limit the number of learning points used for each test point, the speedup can be considerable. However, Section 3.7 emphasized that it is not enough to compute all the distances from all learning points to each test point and only then reject all but the nearest points. We must develop methods that can be used to rapidly locate the nearest points.

The problem of locating the nearest neighbors to a point in space is a well-known problem in computational geometry.[1] Much of the literature is directed toward two- and three-dimensional problems, but considerable work has also been done on the general p-dimensional problem. An important point to note is that if we are looking for *numnn* nearest neighbors, we require only an approximate solution. For example, assume that we have 10 learning points and we search for the three closest points to a test point. Assume also that the correct ordering of the 10 points on the basis of increasing distance from the test point is [7, 3, 4, 6, 10, 9, 1, 2, 5, 8]. Our approximate search might yield Points 7, 4, and 6. The fact that we missed Point 3 is not really crucial for the purposes of kernel regression. The problem of approximate nearest neighbor searching in p-dimensional space has received attention from many researchers. An excellent review of this field is included in an article by Arya, Mount, Netanyahu, and Silverman.[2]

The *quadtree* concept popularized by Hanan Samet[3] can be generalized to p dimensions and provides a very useful data structure for data modeling. A *quadtree* is a hierarchical data structure based on the principle of recursive decomposition of two-dimensional spaces. The quadtree is a particularly useful data structure for applications in computer graphics, computer-aided design, image processing, and many other related fields. The term *octree* is the three-dimensional equivalent of the quadtree and is also used in many related applications (e.g., solid modeling). Samet discusses a number of higher dimension tree structures with particular properties useful for varying applications. For lack of a better term, the term *p-tree* can be used for a particular p-dimensional equivalent of the quadtree and octree. The definition and properties of the p-tree are discussed in Section 4.2, and methods for utilizing the p-tree for efficient kernel regression modeling are suggested in Section 4.4.

4.2 THE p-TREE

The term p-tree is used to define a special data structure that is particularly useful for efficient kernel regression modeling in p-dimensional space. The data structure is a full binary tree with *treeheight* h (a user-specified parameter). A full binary tree is a tree with exactly 2^h leaf cells. Each learning point falls within one of the leaf cells. The *p*-dimensional space is partitioned so that the number of points per cell is approximately constant. In fact, if *nlrn* (the number of learning points) is a multiple of 2^h, then the number of points per leaf cell is exactly constant. An example of a two-dimensional space partitioned in this manner is shown in Figure 4.1.

For this example, *h* (the treeheight) is 3, and therefore the number of leaf cells is 8. The 24 learning points are equally distributed among the eight cells (i.e., three per cell). The associated tree is shown in Figure 4.2. If there had been 25 rather than 24 learning points, then one of the cells would have four points in it and the others would have three.

The *p*-tree can be used to rapidly identify nearby learning points for each test point. For example, assume a test point has a value of (*X*1 *X*2) such that it falls within cell 1 (i.e., the cell MNPO). If we have specified *numnn* (the desired number of nearest neighbors) as 3, then we might use the three points in this cell to estimate *Y* for this test point. But this doesn't necessarily locate the closest three neighbors. For example, if the test point is located near the upper edge of the cell (i.e., near the line MN), then we might also want to consider points in cell 2. For

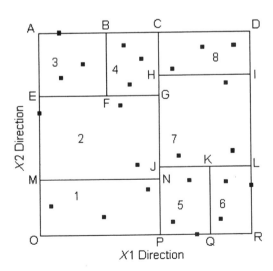

Figure 4.1 Cell distribution in 2D space.

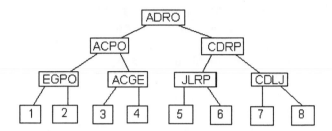

Figure 4.2   Tree representation of Figure 4.1.

the purpose of estimating Y, it is not absolutely essential to identify the exact *numnn* nearest neighbors. The choice of learning points has some influence on predictive accuracy, but if the chosen points are a reasonable selection of nearby points, then the loss of predictive accuracy is small for most real problems.

We can, however, do better than just the points in the cell in which a test point falls. We can build an adjacency matrix that identifies all cells adjoining each cell. As an example, Table 4.1 can be determined directly from Figure 4.1.

Comparing Figure 4.1 and Table 4.1, we can verify that the cell MNPO (i.e., cell 1) has only two adjoining cells: cells 2 and 5. At the other extreme, cell EGNM (i.e., cell 2) has five adjoining cells. If we search not only in the cell in which the test point falls but in all adjoining cells, we will usually find most, if not

TABLE 4.1   Adjacency Matrix for the Tree Shown in Figure 4.2

CELL	Cell Number	Adjacent Cell	Adjacent Cell	Adjacent Cell	Adjacent Cell	Adjacent Cell
MNPO	1	2	5			
EGNM	2	1	3	4	5	7
ABFE	3	2	4			
BCGF	4	2	3	7	8	
JKQP	5	1	2	7	8	
KLRQ	6	5	7			
HILJ	7	2	4	5	6	8
CDIH	8	4	7			

all, of the *numnn* nearest neighbors. We still however, might, miss some close points. For example, if our test point is in the upper right-hand corner of cell 1, we will miss points in the lower left-hand corner of cell 7, even though these might be closer than some of the points located in cells 1, 2, and 5. Extensive testing has shown that neglecting such points results in only minimal loss of predictive accuracy in real problems. In fact, we can often achieve excellent predictive power by only using the points within the cell in which a test point is located.

As the dimensionality of the problem increases, the number of adjoining cells increases exponentially. Searching in all adjacent cells can therefore be quite costly. There is, of course, a middle ground between searching for the *numnn* points only in the test cell or in the test cell plus all adjacent cells. Another parameter (let's call it *numcells*) can be defined as the maximum number of cells to be searched for the *numnn* nearest neighbors. If *numcells* is specified as 1, then only the cell in which the test point falls is used. If *numcells* is specified as 2, then the test cell plus the nearest adjoining cell to the test point will be used. If *numcells* is specified as a value that is greater than the number of adjoining cells plus one, then only the test cell and all the adjoining cells will be used. Clearly, computational time increases as *numcells* increases but only to some limiting value. For the above example, we see from Table 4.1 that the limiting value of *numcells* is 6. For test points in cells 2 and 7, a value of six ensures that for every test point the search for the *numnn* nearest neighbors includes the test cell plus all adjoining cells. Increasing *numcells* beyond 6 will have no effect on the results for this particular *p*-tree.

In summary, the *p*-tree provides a simple method for creating the equivalent of a bandwidth in *p* dimensions. Once the tree has been created, then the particular cell in which a test point falls is located in $O(h)$ time (i.e., the time to locate the cell varies linearly with the *treeheight*). Once the cell has been located, the data structure should then provide immediate access to all learning points in the cell. If *numcells* is greater than one, then the nearest *numcells*–1 adjoining cells can be rapidly identified using the adjacency matrix, and then the learning points in these cells will also be available. For every space examined, use of the p-tree is based on the following procedure:

1. Create a p-tree of height h using the *nlrn* learning points.
2. If *numcells* > 1 create an adjacency matrix.
3. For each of the *ntst* test points:
   a. Locate the cell in the p-tree in which the point falls.
   b. If *numcells* > 1, locate up to *numcells*-1 closest adjacent cells.
   c. Use the *numcells* (or as many cells as are available) to locate the *numnn* nearest learning points.
   d. Use the *numnn* learning points to predict Y for the test point.
4. Use the predicted values of Y to determine MC (the modeling criterion).

4.3 PARTITIONING THE LEARNING DATA SET

When a space is to be examined, the first step is to create a p-tree using the *nlrn* learning data points. Defining h as the *tree-height* of the p-tree, we partition the learning points into the 2^h leaf cells by a series of sorts. For example, consider the two-dimensional learning data in Table 4.2 with *nlrn* = 16. We will use this data to build two p-trees: one with $h = 2$ and the second with $h = 3$. The first tree will have four leaf cells with four learning points per cell, and the second tree will have eight cells with two points per cell.

To create the p-trees, we must first sort the data in either the X_1 or X_2 direction. The choice is arbitrary, but a reasonable heuristic is to start with the longer dimension (i.e., X_1). The first task is to create a sorted list of point indices. For example, after the data is sorted on the basis of X_1, the list will contain the following elements:

LIST = [13, 2, 9, 5, 14, 6, 16, 7, 15, 4, 8, 3, 1, 10, 12, 11];

TABLE 4.2 Learning Data Set

Point	X_1	X_2	Y
1	7.3	0.011	4.5103
2	−4.2	0.032	−2.6614
3	6.1	−0.064	12.7551
4	1.3	0.009	−4.0885
5	−3.7	−0.078	−2.1113
6	−2.8	0.035	0.2670
7	0.6	−0.111	8.5496
8	5.9	−0.001	0.6544
9	−4.2	0.042	−6.6181
10	7.3	−0.072	14.6977
11	9.6	−0.019	12.7309
12	8.2	0.018	10.0081
13	−4.9	0.038	−11.6614
14	−3.2	0.074	−9.9776
15	1.1	−0.091	4.1284
16	−1.6	0.004	2.2795

We see that the first element in the list is 13 because the smallest value of X_1 is at Point 13. The last element is 11 because the largest value of X_1 is at Point 11. Examination of the values of X_1 shows that both Points 2 and 9 are the same (i.e., −4.2). In other words, the sorted list might have been as follows:

LIST = [13, 9, 2, 5, 14, 6, 16, 7, 15, 4, 8, 3, 1, 10, 12, 11];

The only difference in the two lists is that elements 2 and 3 are reversed. Reversals of this type might lead to different results, but both are correct. Depending on the sorting routine used to create the list, either result is possible. We can consider the p-tree as having a root node that points to all the $nlrn$ learning points. This root node is considered as level 0 of the tree. The initial sort creates an ordered list of points, and this list is then used to divide the $nlrn$ points into two subsets. Level 1 of the tree contains two nodes: the *left* and *right subtrees* of the root node. The sorted list is used to make the level 1 partition. The first half of the list (i.e., Points 13, 2, 9, 5, 14, 6, 16, and 7) is placed in the left subtree, and the second half of the list (i.e., Points 15, 4, 8, 3, 1, 10, 12, 11) is included in the right subtree.

At this point we can identify a cell boundary. The first half of the list includes the first eight points, and the second half includes the remaining eight points. The cell boundary should therefore be between the eighth and ninth points in the list. These are Points 7 and 15, and the values of X_1 for these two points are 0.6 and 1.1. If we set a boundary as halfway between these two points, the boundary in the X_1 direction is 0.85. This value is the boundary between the two subtrees of the root node. One might ask the question: What if the two values of X_1 are the same? The boundary would simply take on this value. It is not a problem if points with the same value of X_1 (or any other dimension) are in different cells.

Once the list has been sorted in the X_1 direction, the next step is to sort each half of the list in the X_2 direction. These two sorts are required to create the four cells at level 2. Each sort creates two new subtrees. The points with lower values of X_2 are in the new left subtrees, and the remaining points are in the new right subtrees. The first sort yields the following order: [7, 5, 16, 2, 6, 13, 9, 14]. The values of X_2 for the fourth and fifth points in this list (i.e., Points 2 and 6) are 0.032 and 0.035. Thus the boundary between the two halves of this list is 0.0335. The second sort yields the following order: [15, 10, 3, 11, 8, 4, 1, 12]. Using the same procedure, we see that the values of X_2 for Points 11 and 8 are –0.019 and –0.001, so the boundary is –0.010. Combining the two lists, we replace the original list with the following:

LIST = [7, 5, 16, 2, 6, 13, 9, 14, 15, 10, 3, 11, 8, 4, 1, 12];

At this point the list contains four groups of data points: the first two quarters of the list each includes points with lower values of X_1; the last two quarters contain points with higher values of X_1. The first and third quarters contain points with lower values of X_2, and the second and fourth quarters contain points with higher values of X_2.

We can now build a tree of height $h = 2$. There are four leaf cells in a tree of this height and four learning points in each cell.

TABLE 4.3   Distribution of Learning Points in Tree of Height 2

Cwll	X_1	X_2	Points
1	$-4.9 <= X_1 <= 0.85$	$-0.111 < X_2 <= 0.0335$	7, 5, 16, 2
2	$-4.9 <= X_1 <= 0.85$	$0.0335 < X_2 <= 0.074$	6, 13, 9, 14
3	$0.85 < X_1 <= 9.6$	$-0.111 < X_2 <= -0.010$	15, 10, 3, 11
4	$0.85 < X_1 <= 9.6$	$-0.010 < X_2 <= 0.074$	8, 4, 1, 12

The distribution of points into the cells is summarized in Table 4.3.

Table 4.3 shows that the groups of learning points represented by the lower-valued X_1's are in cells 1 and 2, and the points represented by the higher values are in cells 3 and 4. The lower values of X_2 are in cells 1 and 3, but the upper boundaries of these two cells are different. For this simple example, the first four points in the updated list are in cell 1, the next four are in cell 2, and so on. This distribution of points into cells is shown in dimensionless (X_1, X_2) space in Figure 4.3.

In Figure 4.3 the location of the points are presented in a dimensionless form. Each value of X_1 is transformed to $(X_1 - X_1\_MIN)/Range1,$ and the values of X_2 are transformed in a

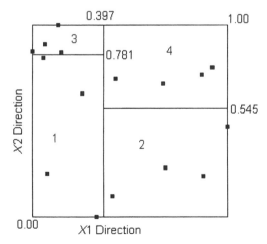

Figure 4.3   Four-cell distribution of learning points.

similar manner. The cell boundaries are also computed in a similar manner. For example, the boundary in the X_1 direction is computed as $(0.85 - -4.9)/14.5 = 0.397$. This cell distribution is presented as a binary tree in Figure 4.4.

To create a p-tree of treeheight 3, we must continue the process a step further. At this point, however, we must decide in which direction to perform the sorts for each of the four cells. Since we have already sorted in both directions, a criterion is needed for deciding the direction for the next sort. A useful heuristic is to sort in the "longest" direction. An examination of Table 3.7 shows that the range of values of X_1 is -4.9 to 9.6 (i.e., $Range1$ = 14.5). We also see that $Range2$ = 0.185 (i.e., -0.111 to 0.074). Because the scaling of the data in the two directions is so different, we must define "longest" in some sort of dimensionless sense. By dividing the lengths of each cell direction by the appropriate range, we determine dimensionless lengths and can then use these numbers to decide on the next sorting direction. The dimensionless lengths are summarized in Table 4.4.

The dimensionless length ratio $R(X_i)$ is defined as the difference between the maximum and minimum values of X_i for each cell divided by the Range of X_i. For example, $R(X_1)$ for cell 3 is simply $(9.6-0.85)/14.5 = 0.603$. We see that for cell 1 the value of $R(X_2) > R(X_1)$ and therefore this cell should be sorted in the X_2 direction. For the remaining cells, the situation is reversed: $R(X_1) > R(X_2)$, so the sorts should be performed in the X_1 direction.

Proceeding to the sorting of the four cells, cell 1 greets us with a pleasant surprise: it is already sorted in the X_2 direction! The four points in this cell (i.e., 7, 5, 16, and 2) are thus divided into two cells with Points 7 and 5 in the left subtree and 16 and

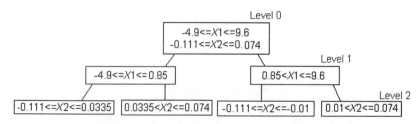

Figure 4.4 Tree of height two based on Figure 4.3.

TABLE 4.4   Cell Boundaries and Dimensionless Lengths R:
(Max – Min)/Range

Cell	X_1-MIN	X_1-MAX	$R(X_1)$	X_2-MIN	X_2-MAX	$R(X_2)$
1	–4.90	0.85	0.397	–0.1110	0.0335	0.781
2	–4.90	0.85	0.397	0.0335	0.0740	0.219
3	0.85	9.60	0.603	–0.1110	–0.0100	0.545
4	0.85	9.60	0.603	–0.0100	0.0740	0.455

2 in the right subtree. Cell 2 (i.e., Points 6, 13, 9, and 14) requires a sort in the X_1 direction. The two lower valued points are 13 and 9, and the higher valued points are 14 and 6. Sorting cell 3 in the X_1 direction puts Points 15 and 3 in the left subtree and points 10 and 11 in the right subtree. Cell 4 (i.e., 8, 4, 1, and 12) requires an X_1 sort. The left subtree points are 4 and 8, and the right subtree points are 1 and 12. The list at this point is as follows:

LIST = [7, 5, 16, 2, 13, 9, 14, 6, 15, 3, 10, 11, 4, 8, 1, 12];

We can summarize the resulting p-tree of height 3 in several ways. The cell division is shown in dimensionless (X_1, X_2) space in Figure 4.5. The cell details are shown in Table 4.5.

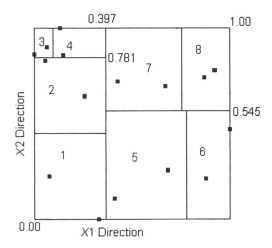

Figure 4.5   Eight-Cell distribution of learning points.

TABLE 4.5 Distribution of Learning Points in Tree of Height 3

Cell	X_1	X_2	Points
1	$-4.9 <= X_1 <= 0.85$	$-0.111 <= X_2 <= -0.041$	7, 5
2	$-4.9 <= X_1 <= 0.85$	$-0.041 < X_2 <= 0.0335$	16, 2
3	$-4.9 < X_1 <= -3.7$	$0.0335 < X_2 <= 0.074$	13, 9
4	$-3.7 < X_1 <= 0.85$	$0.0335 < X_2 <= 0.074$	14, 6
5	$0.85 < X_1 <= 6.7$	$-0.111 < X_2 <= -0.010$	15, 3
6	$6.7 < X_1 <= 9.6$	$-0.111 < X_2 <= -0.010$	10, 11
7	$0.85 < X_1 <= 6.6$	$-0.010 < X_2 <= 0.074$	4, 8
8	$6.6 < X_1 <= 9.6$	$-0.010 < X_2 <= 0.074$	1, 12

We can make some general observations regarding the creation of a p-tree of height h in p-dimensional space using n learning points:

1. Sorts are required to specify the distribution of points at some or all of the h levels of the tree.
2. The maximum number of sorts required to proceed from level $i - 1$ to level i is $2^{(i-1)}$.
3. For all levels, $i <= p$, a different dimension should be chosen for the sort. A useful heuristic is: at every level choose the longest remaining unsorted dimension.
4. For each sort at all levels $i > p$, the dimension with the largest dimensionless length ratio should be chosen. There will only be levels with $i > p$ if $h > p$.
5. The final sorts bring us to level h and are of approximate size $n/2^{(h-1)}$.
6. For the simple case of one-dimensional data (i.e., $p = 1$), only one sort is required regardless of the treeheight h.

The maximum time required for all the sorts can be estimated by assuming that the time to sort a list of n elements is $O(n * \log2(n))$. At the first level a single sort of all n elements is required. If the time to sort all the elements at level 1 of the tree is proportional to $n * \log2(n)$, then at level 2 two sorts of $n/2$ elements each are required (unless $p = 1$). The time for these two sorts would therefore be proportional to $2 * (n/2) * \log2(n/2)$, which is just $n*(\log2(n)-1)$. Similarly, the time required for the four sorts at level 3 would be just $n*(\log2(n)-2)$. We can develop a general equation for the maximum time for all sorts up to level h by summing terms of this form:

$$Sort\_Time \leq C * \sum_{i=1}^{h} n * (\log 2(n) - i + 1) \tag{4.1}$$

In this equation C is a machine-dependent constant. The equation is shown as an inequality because not all sorts must necessarily be made. The equation reduces to the following:

$$Sort\_Times \leq C*n*(h*\log2(n) - (h - 1)*h/2) \tag{4.2}$$

This equations shows that the upper limit for the time to perform all the sorts at all the levels is less than h times the time for a single sort of all n elements. For example, if $n = 8192$, then $\log2(n) = 13$. If we create a p-tree with $h = 8$, the term inside the parentheses (i.e., $h*\log2(n) - (h - 1)*h/2$) is $8*13 - 7*8/2 = 76$. In other words, the maximum total sorting time is no more than $76/13 = 5.8$ times the time required for a single sort of all 8192 elements. The actual sort time can be significantly less than this limit if some of the sorts at some of the levels are not required.

4.4   USING THE *p*-TREE

The purpose of the p-tree is to rapidly identify nearest neighbors to test points. The nearest neighbors are from the learning data set and are used to estimate the values of Y for the test points. Information is stored in the inner nodes of the tree as well as in the leaf cells. The tree is used to determine the leaf cell of each test points. For a tree of height h, a series of h conditional statements are required to locate the appropriate leaf cell.

To demonstrate the procedure, consider the p-tree developed in Section 4.3. The cell distribution is shown in Figure 4.5 and Table 4.5. This same information is presented as a tree in Figure 4.6.

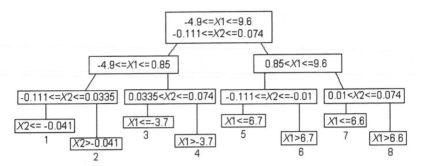

Figure 4.6 Tree representation of Figure 4.4.

Consider the test points listed in Table 4.6.

We can follow the progress of each point down the tree. For example, Point 1 passes the first set of conditions in the root node (which is considered as level 0). This point is therefore accepted as a test point within the range of the learning data set. The value of $X1 = -0.5$ directs this point into the level 1 left subtree of the root node. The value of $X2 = 0.063$ directs the point into the level 2 right subtree, and the final test (i.e., $X1 > -3.7$) directs the point into the level 3 right subtree, which is leaf cell 4.

Point 5 presents a different situation: failure is noted on the first test (i.e., $X1 = 9.7$ is *out-of-range* of all the learning data values of $X1$. It is good practice not to attempt to estimate values of Y for test points that are *out-of-range*. Since the surface being modeled might be highly nonlinear, extrapolation may lead to large errors. Conceptually, we are using historical data in an attempt to build a model. Therefore, points that are out-of-range of the historical database might behave in a very different manner than the historical points. Table 4.7 summarizes the cell distribution of all five test points.

TABLE 4.6   Test Data Set

Point	X1	X2	Y
1	−0.5	0.063	−3.008
2	−4.3	−0.011	−3.531
3	0.8	−0.062	6.691
4	6.4	0.025	−1.010
5	9.7	−0.024	5.169

TABLE 4.7   Distribution of Test Points into the *p*-tree Leaf Cells

Point	X1	X2	Level 1	Level 2	Level 3	Leaf Cell
1	−0.5	0.063	left	right	right	4
2	−4.3	−0.011	left	left	right	2
3	0.8	−0.062	left	left	left	1
4	6.4	0.025	right	right	left	7
5	9.7	−0.024	fails			0

It is reasonable to discard all points exhibiting *out-of-range* values in any direction. Therefore, Point 5 is discarded, and only Points 1 through 4 are used to evaluate the space.

Additional information must be stored in the leaf cells. In Section 4.2 the concept of an *adjacency matrix* was introduced. The adjacency matrix locates leaf cells that are adjacent to the test point cells. If, for example, we wish to find the single nearest adjacent cell to a test point, we would first use the matrix to find all adjacent cells and then determine which of these cells is the closest. The information contained in each leaf cell must be sufficient to determine the closest cell.

The determination of the nearest cell is affected by the definition of the term "nearest." Do we define "nearest" as the cell having the nearest point on any surface of the cell to the test point? Alternatively, we might choose the cell with the nearest center point to the test point. Clearly, all distances must be measured in dimensionless units. In Figure 4.7, test point j is used to demonstrate the type of calculation required to determine the nearest cell. Points s1, s2, s6, s7, and s8 are the nearest points to the surfaces of the cells adjacent to cell 5 (i.e., the cell in which test point j is located). Points m1, m2, m6, m7, and m8 are the midpoints of these cells. If we use the first definition, then all the distances from Point j to s1, s2, s6, s7, and s8 would have to be measured, and in this way the closest cell is located. From the figure we see that cell 7 is closest in this sense. Using the second definition, we would use distances from Point j to m1, m2, m6, m7, and m8. The two definitions might lead to different conclusions, but this difference is not particularly important in as much as we are only interested in an approximate solution to the nearest neighbor problem. As the dimensionality

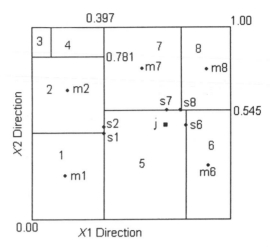

Figure 4.7   Cell distances from test point j.

of the model increases, the task of finding the nearest point on the surface of each adjacent cell becomes more difficult. Use of the midpoints is less computationally complex and is therefore the recommended solution.

Any definition of "nearest" will not necessarily locate all the nearest points in adjacent cells. For example, referring to Figure 4.7, we see that both definitions lead to the conclusion that cell 7 is the closest cell to Point j. If, for example, the learning points in cell 7 are far from Point s7 and some points in either cell 6 or 8 are near Points s6 and s8, these points will be missed if only one adjacent cell is searched. Since only an approximate solution to the nearest neighbor problem is required, this loss of accuracy is usually acceptable. If the number of adjacent cells to be used (i.e., *numcells* − 1) in the search for nearest neighbors is greater than or equal to the number of adjacent cells, then there is no need to compute distances as all adjacent cells are used.

The p-tree is used to make predictions of Y for the test points. A variety of algorithmic choices were introduced in Chapter 3. We can use either the *Order 0, Order 1,* or *Order 2* algorithms. In addition, we can consider the different options introduced in Section 3.5: the *static, growing,* or *moving window* learning data set. Using the $h = 3$ p-tree, the results for a variety of cases are shown in Tables 4.8 and 4.9. In these tables,

Table 4.8   Variance Reduction Using a Static
Learning Data Set

Case	Order	Numnn	Numcells	VR
1	0	2	1	59.6
2	0	3	2	65.4
3	0	4	2	21.8
4	0	4	6	72.6
5	1	4	2	25.0
6	1	4	3	24.9
7	1	5	3	49.3
8	1	6	3	53.8
9	1	6	6	−21.7

order refers to the algorithm, *numnn* is the number of nearest neighbors used to estimate Y, and *numcells* is the number of cells searched to find the nearest neighbors. If *numcells* = 1, then only the test cell learning points are considered. The resulting values of VR are variance reduction as defined by Eq. 1.12.

It is instructive to go through one of the calculations in detail. Consider the values of VR for Case 1 in which the results are the same in both tables. From Table 4.7 we see that each of the first four test data points falls into a different leaf cell (i.e., cell 4, cell 2, cell 1, cell 7). Using the Order 0 Algorithm with

Table 4.9   Variance Reduction Using a Growing
Learning Data Set

Case	Order	Numnn	Numcells	VR
1	0	2	1	59.6
2	0	3	2	65.4
3	0	4	2	18.7
4	0	4	6	72.6
5	1	4	2	25.0
6	1	4	3	26.4
7	1	5	3	80.4
8	1	6	3	74.0
9	1	6	6	−40.3

unit weighting, we see that the estimated values of Y are the average values of the Two learning points (numnn = 2 for this case) located in the test cell (numcells = 1 for this case). From Table 4.7 we see that cell 4 includes Points 14 and 6, cell 2 includes Points 16 and 2, cell 1 includes Points 7 and 5, and cell 7 includes Points 4 and 8. The values of Y for these points are found in Table 4.2. The computed values are included in Table 4.10. For example, the computed value of $Ycalc$ for the first test point is −4.855, which is the average of Y for Points 14 and 6.

Equation 1.12 is used to determine VR:

$$\text{VR} = 100 * \left(1 - \frac{\sum_{i=1}^{ntst} (Y(i) - Ycalc(i))^2}{\sum_{i=1}^{ntst} (Y(i) - Yavg)^2} \right) \tag{1.12}$$

Since $Yavg$ is −0.858/4 = −0.2145, the term in the denominator is 67.121. The numerator term is 27.12, so the value of VR is 100* (1 − 27.12/67.121) = 59.6. Notice that the value determined using the *growing* option (i.e., in Table 4.9) is exactly the same for Case 1. The explanation for this result is quite simple. Each of the four test points falls into a different cell, so they do not affect the cell composition for the subsequent points. The first difference noted between the results in the two tables is for Case 3.

On the basis of these results, it is difficult to come to any conclusions regarding the choice of parameters. There are just too few learning and test points to make serious generalizations. The results in these tables do, however, provide a useful benchmark

Table 4.10   Case 1 Details for the Four in-Range Test Points

Test Point	Y	Ycalc	(Y − Ycalc)²
1	−3.008	−4.855	3.413
2	−3.531	−0.191	11.159
3	6.691	3.219	12.054
4	−1.010	−1.717	0.493
Total	−0.858	Not used	27.119

for developing kernel regression software. In Chapter 5 results are presented for cases in which a large number of both learning and test points are used. The results in Chapter 5 are much more statistically meaningful.

4.5 TIME WEIGHTING THE DATA

When dealing with time series problems, the *time dimension* is often important. For example, when modeling financial data, it is reasonable to assume that more recent data points are more relevant than data from an earlier time period. A straightforward method for attaching greater importance to more recent points is to *time weight* the data. The simple exponential kernel proposed in Section 3.1 (i.e., Eq. 3.1) can be modified:

$$w(x_i, x_j, t, k, \alpha) = e^{-(kD_{ij}^2 + \alpha t)} \tag{4.3}$$

In this equation, t is the time difference between the times associated with the jth test point and the ith learning point. The time constant α is user defined and can be based on the analyst's feeling regarding the *information half-life* of data. For example, if time is measured in units of days and the analyst feels that information from a year ago (i.e., 365 days ago) should receive half the weight of current information, then α would be computed as follows: $0.5 = \exp(-365\alpha)$, which results in a value of $\alpha = 0.0019$.

In Section 4.4 a number of variations on the basic kernel regression paradigm were discussed. Besides the *fit order* (i.e., 0, 1, or 2), the *static, moving window,* and *growing* options were discussed. Use of the parameter *numcells* specifies the number of cells to be searched for the *numnn* nearest neighbors. When time weighting is used, one must decide how "nearest" is defined. Up to this point we have used D_{ij}^2 as the basis for determining the nearest neighbors. However, Eq. (4.3) offers an alternative based upon $kD_{ij}^2 + \alpha t$. Use of this alternative form complicates the entire process, and it is not clear that this added complexity leads to better models. The suggested procedure is to continue to use $D_{ij}{}^2$ to determine the nearest neighbors, but then use Eq. (4.4) to compute the weights.

When $k = 0$ and $\alpha = 0$ (i.e., all points are equally weighted), and if *numcells* = 1 (i.e., only the test cell is used for determining the nearest neighbors), then a very fast analysis can be obtained. If *numnn* is exactly equal to the number of points in the test cell then all points are equally weighted, and thus one surface is used within the cell regardless of the exact location of the test point. A single surface is generated regardless of the fit order. The only difference is the actual surface. If the order is zero, then the surface is a constant equal to the average value of Y of the learning points in the cell; if order is one, then the surface is a hyperplane; and if order is two, then the surface is a second order multinomial. If the *moving window* or *growing* options are used, then the surfaces must be recalculated but only when a learning point has been added or removed from the test cell since the last visit to that cell.

At first glance, it would appear that a value of α other than zero would eliminate the possibility of a very fast analysis even if $k = 0$. Since the weights would be different for every test point, it would appear that the surfaces would have to be recomputed. Fortunately, this is wrong! By using an exponentially decaying time weighting function, the relative weights of all the learning points in the cell remain constant. Thus the surface remains constant. Only if points have been added or removed from a test cell is there a need to recompute the surface.

In summary, time weighting of data can be accomplished by adding a time-dependent term to the exponent of the weight function (i.e., Eq. 4.4). The creation of the p-tree should not be dependent on this term. If $k = 0$, then the relative weights of a set of learning points does not change even though α is not zero. This fact can be utilized to perform rapid analyses even if time weighting is used.

4.6 MULTISTAGE MODELING

The concept of multistage modeling was first introduced in Section 2.3 where the *curse of dimensionality* was discussed.

Briefly, for very complicated systems such as financial markets, one would expect that a large number of variables would be required to develop useful models. However, the data density decreases exponentially with the number of variables included in the model. In other words, a large number of variables are needed, but often there are not enough data to develop a single high-dimensional model. One useful strategy to overcome this problem is to develop multistage models.

The idea behind multistage modeling is quite simple. If we can develop several lower dimensional models that do have some predictive power, then perhaps these models can be combined in such a way that the combined model has even greater predictive power. As an example of such a model, consider Figure 4.8.

This figure presents a final model that is created using five submodels. Models 1, 2, and 3 are first-stage models. Models 4 and 5 are second-stage models, and the Final Model is a third-stage model. Model 1 is a three-dimensional model that takes three of the original candidate predictors as its inputs. Model 2 is four dimensional, and Model 3 is two dimensional. Note that the variable X_{72} is an input variable to both Models 2 and 3. Model 4 accepts as inputs the outputs of both Models 1 and 2 plus an original candidate predictor X_{41}. Model 5 accepts only the outputs from Model 2 and 3, and the Final Model takes as its inputs the outputs from the two second-stage models (i.e., Models 4 and 5). In this particular example, nine of the original candidate predictors are used to generate predictions from the Final Model. The highest dimensionality of any of the models is only four (i.e., in Model 2). In general:

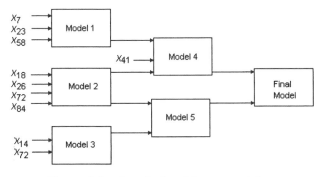

Figure 4.8 A typical multistage model.

1. The first-stage models can only take original candidate predictors as inputs.

2. Higher stage models may take outputs from lower stages as well as original candidate predictors as inputs. (See, for example, Model 4.)

3. Inputs to one model can also be inputs to other models (for example, X_{72}).

4. Outputs of one model can be inputs to more than one other model in a later stage. (For example, the output of Model 2 is an input to both Models 4 and 5.)

5. The maximum number of stages can be determined using some sort of heuristic rule. For example, every model must exhibit some specified improvement over the lower level subspaces of the model. For the above example, if the specified improvement in the modeling criterion is δ, then the Final Model would have had to exhibit at least an improvement of δ over both Models 4 and 5. Similarly, Model 1 would have had to exhibit such an improvement over models based on the subspaces (X_7, X_{23}), (X_7, X_{58}), and (X_{23}, X_{58}).

When can multistage modeling be expected to succeed in generating superior models? The basic requirement is that the correlation between model outputs is not very high. If two models are highly correlated, combining them does not yield significant additional information. Use of the δ criterion will automatically weed out highly correlated combinations of models. For well-defined systems where there is really one true low dimensional model and very little noise, multistage modeling should be ineffective. However, financial markets are complex and noisy, and often several models can be determined which do provide some useful information and yet are not highly correlated. For such cases, multistage modeling is a worthwhile technique.

4.7 THE SEARCH ENGINE

High-performance modeling software based on kernel regression requires a search engine. The function of the search engine is to choose the spaces to be examined. In Section 3.6 a strategy for conducting such a search was discussed, and the need for such a strategy was illustrated using Table 3.10. In that table, the total number of spaces are shown as a function of ncp (the number of candidate predictors) and $dmax$ (the maximum dimensionality of all spaces considered). The total number of spaces increases rapidly as a function of these two parameters. For values of $ncp = 200$ and $dmax = 5$, the total number of spaces is over 2.6 billion!

The search engine provides the analyst with a vehicle for controlling the total number of spaces to be examined. In Section 3.6 the concept of a parameter $num\_survivors(d)$ was introduced. This parameter is the upper limit of the number of spaces that may survive from an analysis of spaces of dimensionality d. This parameter is independent of the choice of modeling criterion. For each space examined, we can determine a value of $MC(i)$, which is the modeling criterion for the ith space. The most simple-minded approach to the design of a search engine is to sort the values of $MC(i)$ for dimensionality d. Then select the top $num\_survivors(d)$ as the survivors for this dimensionality. These survivors then form the basis for creating spaces of dimensionality $d + 1$.

A major problem in modeling when ncp (the number of candidate predictors) is large is the problem of overfitting. One would like to avoid choosing a high-dimensional model if a lower dimensional model is comparable in performance. Usually, some sort of penalty function is used which punishes higher dimensional models. (One example is the well-known Akaikhe Information Criterion.[4]) A simple approach to this problem is discussed in Section 3.6: use of a parameter δ. The δ parameter is defined as the minimum improvement in MC that a space must exhibit to be considered for survival. All spaces examined at dimensionality d (if $d > 1$) must have a value of MC at least δ greater than its subspaces of dimensionality $d - 1$.

It is inefficient to save all results at dimensionality d and only then sort the results to determine the $num\_survivors(d)$. A more efficient approach to this problem is to use a linked list ordered on the basis of MC. For each new space examined, if it passes the δ criterion, then it should be inserted into the linked list if the list contains less than $num\_survivors(d)$. If the list contains exactly *num_survivors(d)*, then it should be inserted into the list if its value of *MC* is high enough to qualify. The last member of the list would then have to be discarded. For greatest efficiency, the list should be entered from the low end rather than the high end. If the number of spaces to be examined at dimensionality d is much larger than *num_survivors(d)*, then most spaces will be rejected. If one enters the list from the low end, then most spaces will immediately fail to qualify.

The parameter *num_survivors(d)* is an upper limit. If the δ criterion is used, then the number of survivors might be less than *num_survivors(d)*. In fact, the number of survivors may even be zero. At this point the analysis would cease even if the dimensionality d were less than the specified value of *dmax*. This is perfectly reasonable. The specified value of *dmax* should be considered only as an upper limit. The search engine should have the capability of terminating a search when it appears to be pointless to continue.

Another interesting feature of a search engine is the ability to force in some selected candidate predictors. If, for example, it becomes clear that X_1 is so powerful that it definitely belongs in the model, it makes sense to start the analysis from dimensionality $d = 2$ and only look at spaces paired with X_1 (i.e., (X_1, X_2), (X_1, X_3), etc.). To generalize this concept, the parameters *nf* (number of forced subspaces) and *dimf* (dimensionality of the forced subspaces) must be specifiable. The user would then have to be able to specify the actual *nf* subspaces. For the above example, *nf* = 1, *dimf* = 1, and the forced subspace is X_1. As a further example, assume we wish to initiate the analysis from $d = 3$ based on the 2D subspaces (X_1, X_4), (X_1, X_7) and (X_2, X_9). For this example, *nf* = 3 and *dimf* = 2.

4.8 COMPUTATIONAL COMPLEXITY

In Section 3.7 the subject of the timing associated with the modeling process was considered. Equation (3.44) states that the time required to develop a set of models is approximately the product of the number of spaces examined times the average time required per space:

$$T_{total} = S_{total} * T_{avg} \qquad (4.4)$$

This equation is based on the assumption that the computer being used for the calculations exploits a single processor. As of early 1999, most computers included only one processor, but personal computers with two, four, and even eight processors were already commercially available. Parallel processing can theoretically reduce the value of T_{total} by a factor of up to N where N is the number of available processors. Parallel processing as it applies to the process described in this chapter is considered in greater detail in Section 4.9.

The use of a p-tree as the basis for kernel regression modeling was introduced in Section 4.2. In this section the computational complexity associated with this modeling technique is considered in greater detail. The average time per space T_{avg} can be considered as consisting of two basic components: the preparation time and the run time:

$$T_{avg} = T_{prep} + T_{run} \qquad (4.5)$$

The preparation time T_{prep} consists of three components: the Sort_Time, the time required to develop an adjacency matrix if adjoining cells are to be used to locate nearest neighbors, and some bookkeeping required for all the nodes in the tree:

$$T_{prep} = Sort\_Time + T_{adj} + T_{book} \qquad (4.6)$$

In Section 4.3 an upper limit for Sort_Time was developed. Using standard sort routines such as heap_sort or quick_sort, we can achieve $O(n*\log2(n))$ complexity. For a tree of height h, it was shown that the sum of sorts at each level yields a maximum sort time as shown in Eq. (4.2).

$$Sort\_Time \le C* n* (h* \log 2(n) - (h - 1)* h/2) \qquad (4.2)$$

In this equation n is the number of learning points used to create the tree, h is the treeheight, and C is a machine-dependent constant. This equation for the Sort_Time represents an upper limit. A new sort is required only when the ordering of a particular sort is based on a dimension different from the sort dimension at the preceding level. For example, for one-dimensional models, only a single sort is required because all orderings at lower levels in the tree are based on the same dimension.

The second term in Eq. (4.6), T_{adj}, is the time required to generate the adjacency matrix and is only nonzero if this matrix is required. If only the cell in which a test point falls is used, an adjacency matrix is not needed. However, if an adjacency matrix is required, this term is important, and for large values of h it can be the dominant term. T_{adj} is proportional to the square of the number of leaf cells. Since the number of leaf cells is equal to 2^h, T_{adj} is proportional to 2^{2h}:

$$T_{adj} = C_{adj}\, 2^{2h} \qquad (4.7)$$

In other words, increasing the value of h by one increases the value of T_{adj} by a factor of four. If the number of learning points per leaf cell is held constant, then we can relate h to n:

$$Points\_per\_cell = n/2^h \qquad (4.8)$$

Values of n for different combinations of h and ppc (*Points_per_cell*) are shown in Table 4.11.

Multiplying Eq. (4.8) by 2^h and then taking log2 of both sides of the equation, we can see that if ppc is held constant, the treeheight h increases as $\log2(n)$. Substituting this into Eq. (4.7), we see that T_{adj} is proportional to $2^{\log2(n)}$, which increases at a faster rate than $n*\log2(n)$. We can therefore conclude that

TABLE 4.11   Values of n for Combinations of h and *Points_per_cell*

h	10 ppc	20 ppc	40 ppc	80 ppc	160 ppc
5	320	640	1280	2560	5120
6	640	1280	2560	5120	10240
7	1280	2560	5120	10240	20480
8	2560	5120	10240	20480	40960
9	5120	10240	20480	40960	81920
10	10240	20480	40960	81920	163840
11	20480	40960	81920	163840	327680
12	40960	81920	163840	327680	655360
13	81920	163840	327680	655360	1310720

if the *Points_per_cell* is held constant, then for large n the *Sort_Time* becomes small compared to T_{adj}. It should be mentioned once again that if the search for nearest neighbors is limited to only the test cell data points, then there is no need to determine an adjacency matrix and thus T_{adj} is zero.

The final term in Eq. (4.6), T_{book}, is proportional to the number of nodes in the tree:

$$T_{book} = C_{book} (2*2^h - 1) \qquad (4.9)$$

Based on the same reasoning, if *Points_per_cell* is held constant, then T_{book} is proportional to $2^{\log 2(n)}$, which increases at a slower rate than both T_{adj} and *Sort_Time*. (A doubling of n requires that h be increased by one in order to maintain a constant value of *ppc*. Increasing h by one causes an increase in T_{book} by about a factor of 2, an increase in the *Sort_Time* by a factor slightly greater than 2, and an increase in T_{adj} by a factor of 4.) Thus for large values of n, the dominating term is T_{adj} (if it is nonzero). Timing studies in which T_{prep} is measured are included in Chapter 5.

The run time T_{run} is proportional to the number of test points *ntst:*

$$T_{run} = C_{run} ntst \qquad (4.10)$$

The constant C_{run} (run time per test point) is dependent on many kernel regression parameters. The value of C_{run} can be so small that T_{run} is negligible compared to T_{prep}. Alternatively, it can be the dominating term. The main components that enter into the time per test point are C_{cell} (the time to determine in which cell a test point resides), C_{search} (the search time required to find the *numnn* nearest neighbors), C_{weight} (the time to determine the weights for each of the *numnn* nearest neighbors), and C_{solve} (the time required to solve for the value of *ycalc(i)*):

$$C_{run} = C_{cell} + C_{search} + C_{weight} + C_{solve} \qquad (4.11)$$

The term C_{cell} is proportional to h (the height of the tree). This time is very small and is essentially the time to perform h *if* statements. The term C_{search} may be zero, or it may be an important component of C_{run}. It is zero if only the learning points in the test cell are used to determine *ycalc(i)*. (If only these learning points are used, then there is no need to perform a search for nearest neighbors.) If, however, a search is to be performed, C_{search} is proportional to *numcells * Points_per_cell* (the product of the number of cells to be included in the search and the number of points per cell). It is also dependent on *numnn* because a list of this length must be managed as part of the search procedure. The term C_{weight} is proportional to *numnn* (the number of nearest neighbors) unless all points are equally weighted. If all points are equally weighted, then this term is zero.

The final term is Eq. (4.11) C_{solve} consists of several steps: first, the terms of a matrix equation (Eq. 3.8) must be determined. The second step is solving the set of equations for the coefficients of the surface (i.e., the A_j's in Eq. 3.7). The final step is to determine the value of *ycalc* using Eq. (3.7) for the test point. The size of the matrix equation is dependent on the dimensionality of the space and the order of the algorithm used to model the space. If *Order 0* is used, the matrix equation is just a single equation with one unknown; if *Order 1* is used, the number of equations is *dim* + 1 (where *dim* is the dimensionality of the space; and if *Order 2* is used, the number of equations is $1 + dim + (dim + 1)*dim/2$. The time to solve a set of linear equations is approximately proportional to the number of equations raised to the third power (i.e., $num\_equations^3$). The final step (solving for the value of *y* from Eq. 3.8) is small compared to the time required to solve the equations and is proportional to *num_equations*. If only the learning points in the test cell are used, and if all the points are equally weighted, then the process need not be repeated once the A_j's have been determined. (If the *growing* or *moving* options are used, then once a learning point is added or removed from a particular cell, the A_j's would have to be recomputed when a new test point falls into the cell.)

From the preceding discussion, it can be concluded that the conditions to make C_{run} very small are as follows:

1. Use only the learning points within the test cell.
2. Weight all points equally.

3. Save the coefficients of Eq. (3.8) (i.e., the A_j's).

4. When learning points have been added or removed from a cell, recompute the A_j's only after a new test point is located in the cell.

When such a strategy is used, the running time T_{run} is usually small compared to T_{prep}. In addition, this strategy reduces T_{prep} because there is no need to determine an adjacency matrix. For cases in which greater accuracy is required, a search for nearest neighbors is usually made, and some weighting (either distance or time or both) might be desirable. For such cases, the running time is usually the dominating term in Eq. (4.5). Some timing comparisons are presented in Chapter 5.

4.9 PARALLEL PROCESSING

The discussions of timing considerations in Sections 3.7 and 4.8 were based on the assumption of a single processor machine. Single processor architecture has been the dominant computer model since John Von Neumann originally proposed it in the 1940s. However, one doesn't have to be clairvoyant to realize that the single processor machines days of domination are numbered. Already one can purchase, at modest prices, personal computers with two, four, and eight processors on a single board, and soon several processors will fit on a single chip.[5] Massively parallel machines are also available but at greater cost. The problem with exploiting parallel processing is that software must be written to utilize the potential of these machines.

There are many competing parallel architectures, and there are software solutions for each type. At the lower end of parallel computing, currently the most well-known types of architecture are SMP (symmetric multiprocessors), NUMA (nonuniform memory access), and clusters.[6] In this section the emphasis will be on SMP machines because these types of machines can easily be applied to speeding up the modeling process as described in this chapter.

An important feature of SMP machines is that memory is shared. Each processor works independently, and each accesses the same memory. There is an inherent danger in this model. If, for example, one processor is reading from a particular memory location while another processor is attempting to write to the same location, the results will be different depending on which command is executed first. The use of the different processors must be synchronized so that these types of conflicts are avoided and the results are correct. In this section we consider how one would go about programming an SMP machine to speed up the modeling process as described in the previous sections of this chapter.

In Section 3.7 the subject of the timing associated with the modeling process using a single processor was considered. Equation (3.44) states that the time required to develop a set of models is approximately the product of the number of spaces examined times the average time required per space:

$$T_{total} = S_{total} * T_{avg} \tag{3.44}$$

Parallel processing can theoretically reduce the value of T_{total} by a factor of up to N where N is the number of available processors:

$$T_{total} >= S_{total} * T_{avg} / N \tag{4.12}$$

The most obvious strategy for parallelizing the modeling process is to analyze a separate space on each processor. The partitioning of the learning data is discussed in Section 4.3. Each processor would access the same raw data because this data remains constant. However, there is a need to allocate some separate memory for each processor. For example, a separate sorted list of pointers to the learning data records is required for each processor. In addition, separate trees with storage of data in each node of the tree are required for each processor. When analyzing the test data, intermediate results must be saved for each processor separately. The simplest model for M_{total} (total memory required) is:

$$M_{total} = M_{shared} + N * M_{processor} \tag{4.13}$$

In this equation M_{shared} is the shared memory requirement, and $M_{processor}$ is the memory required for each processor. We see that as the number of processors increases, the total amount of memory required increases as $O(N)$. If the value of M_{total} exceeds the available memory, paging begins to occur and one would then expect a degradation in performance. In other words, for a given problem size and available memory, there is a value of N that optimizes performance. Going beyond this point, one can expect to see at first a stagnation in T_{total} and eventually an increase in T_{total} as N is increased further.

Modeling financial markets is an application that can be extremely computer intensive. For users of the modeling technique described in this chapter, the availability of parallel computing hardware presents the analyst with an opportunity to speed up the analysis significantly. The software problems associated with exploiting parallel computing hardware for this application are not particularly difficult because the modeling process lends itself to this mode of computing.

NOTES

1. F. P. Preperata and M. I. Shamos, *Computational Geometry: An Introduction* (New York: Springer-Verlag, 1985).

2. S. Arya, D. Mount, N. Netanyahu, and R. Silverman, "An Optimal Algorithm for Approximate Nearest Neighbor Searching," *Symposium on Discrete Algorithms,* Chapter 63 (New York: Springer-Verlag, 1994).

3. H. Samet, *The Design and Analysis of Spatial Data Structures* (Reading, Mass: Addison Wesley Longman, 1990).

4. H. Akaikhe, "Statistical Predictor Identification," *Ann. Inst. Stat. Math,* 22 (1970): 203–217.

5. D. Culler and J. P. Singh, *Parallel Computer Architecture; A Hardware/Software Approach* (San Francisco, CA: Morgan Kaufmann Publishers Inc., 1999).

6. G. F. Pfister, *In Search of Clusters; the Ongoing Battle in Lowly Parallel Computing* (Englewood Cliffs, NJ: Prentice Hall, 1998).

5

KERNEL REGRESSION
SOFTWARE PERFORMANCE

5.1 SOFTWARE EVALUATION

In Chapter 4 a number of parameters and options for performing kernel regression analyses were introduced. In this chapter the effects of these parameters and options on the software performance are discussed. The software performance can be considered from several points of view:

1. The time required to perform the calculations.
2. The "success" of the software in locating good models.

Timing statistics form the basis for estimating the time required to perform subsequent analyses and are easily obtained. The task of measuring success is less clear-cut. If, for example, real financial data is used and no acceptable model is located, what has been proven? Is the kernel regression methodology or the software at fault, or is there simply no significant relationship between the candidate predictors and the variable being modeled? When evaluating modeling software, the clearest answers are obtained using artificial data with known signal-to-noise ratios.

The suggested procedure is to start out with a Y variable (i.e., the target variable that is being modeled), which is generated from a known model based on some of the candidate predictors as well as a measured amount of random noise. Once the

105

data has been generated, the software can be used and the results can be interpreted exactly. If, for example, a data series is generated which is 10 percent signal and 90 percent noise, then a measured value of VR (Variance Reduction) of 7 percent can be interpreted. The process is 70 percent efficient: it has managed to capture 70 percent of the available signal. By varying the parameters and observing the changes in VR and compute time, the analyst can rapidly develop a feeling for the effect of the various parameters on the software performance.

An important component for performing such analyses is a vector-oriented computer language that can be used to rapidly generate data files. Such languages as APL and MATLAB can easily be used to accomplish this task. My personal preference is for a language called TIMES.[1] The following TIMES code (Figure 5.1) is easily understandable and is used to generate an artificial data set of 15,000 records and 17 columns of data. The first 10 columns are normally distributed random variables. The eleventh column is a pure three-dimensional nonlinear function based on $X2, X5$, and $X9$ (i.e., columns 2, 5, and 9). The next four columns are the eleventh column plus increasing levels of noise. The noise component in column 12 amounts to 50 percent of the total variance in this column. In columns 13, 14, and 15 the noise components are 75, 90, and 95 percent. The kernel regression method does not impose distribution requirements on the candidate predictors. To illustrate this point, the mean of $X2$ is increased by 10, the scale of $X5$ is increased by a factor of 1000, and the scale of $X9$ is decreased by a factor of 1000.

The derivation of the equation used in this program to generate the noise terms is based on the following relationship:

$$\sigma^2_{sig+noise} = \sigma^2_{sig} + \sigma^2_{noise} \tag{5.1}$$

Equation (5.1) requires an assumption that the signal and noise are uncorrelated. Since the noise vector is created using a random number generator, this assumption is valid. Multiplying a random number vector by a constant C creates the noise vector. The constant C can be related to MAXVR (the maximum expected Variance Reduction):

$$\text{MAXVR} = 100 * (1 - \sigma^2_{\text{noise}}/\sigma^2_{\text{sig+noise}}) \qquad (5.2)$$

Solving for σ^2_{noise}:

$$\sigma^2_{\text{noise}} = \sigma^2_{\text{sig}} (1 - \text{MAXVR}/100)/(\text{MAXVR}/100) \qquad (5.3)$$

Since σ_{noise} is equal to C σ_{random}, the value of C is obtained from Eq. (5.3):

$$C = \frac{\sigma_{sig}}{\sigma_{random}} \sqrt{(1 - MAXVR/100)/(MAXVR/100)} \qquad (5.4)$$

In Figure 5.1 the value of σ_{sig} is obtained by computing the standard deviation of the signal column (i.e., `sy = sigma (a[:m + 1])`). The value of σ_{random} is obtained in a similar manner.

The data file created in Figure 5.1 is used as a testbed in the following sections. This file is created using a signal column based on one nonlinear function. Therefore, the results are only strictly valid for this one particular function. However, the purpose of this chapter is to give the reader a qualitative understanding of the impact of the various options and parameters that affect the kernel regression analysis. This nonlinear function adequately serves this intended purpose. The function is derived as the sum of three components: y1, y2, and y3. All three components are functions of a single random variable and are composed of a Gaussian peak at –0.5 and a Gaussian valley at 0.5 (prior to translation or scaling). The average value of each component is therefore close to zero, and so the number of positive and negative signals (i.e., *y1+y2+y3* in Figure 5.1) should be approximately the same. Since the average value of the noise components is also close to zero, the number of positive and negative values in the columns corrupted with noise should also be approximately the same. When one is modeling real financial data, the typical *Y* variables are changes in some quantity and therefore tend to also have a mean value close to zero. One measure of the "goodness" of a model is *FracSS* (i.e., Fraction Same Sign), which is the fraction of the test records in which the predicted and actual values of *y* have the same sign. By having a

```
// TIMES program to generate artificial data.
proc main() {
  n = 15000;     // number of records
  m = 10;        // number of candidate predictors

  a = reshape([n,m+7],0);   // creates an n by m+7 zero filled matrix

  a[:1..m] = grnum(n*m);    // fills columns 1 thru m with Gaussian
                            // random noise  (mean 0, sigma 1)

  y1=exp(-((a[:2] + 0.5)^2)) - exp(-((a[:2] - 0.5)^2));
  y2=exp(-((a[:5] + 0.5)^2)) - exp(-((a[:5] - 0.5)^2));
  y3=exp(-((a[:9] + 0.5)^2)) - exp(-((a[:9] - 0.5)^2));

  a[:2] = a[:2] + 10;       // increase all values by 10
  a[:5] = 1000*a[:5];       // scale values up by a factor of 1000
  a[:9] = a[:9] / 1000;     // scale values down by a factor of 1000

  a[:m+1]=y1+y2+y3;  // The pure signal is a 3D non-linear function

  random=rnum(n);           // rnum generates random numbers from -1 to 1
  sran=sigma(random);       // sigma returns standard deviation
  sy=sigma(a[:m+1]);        // sigma of the pure signal

  a[:m+2]=a[:m+1]+(sy/sran)*sqrt(0.50/0.50)*random;    // maxvr=50
  a[:m+3]=a[:m+1]+(sy/sran)*sqrt(0.75/0.25)*random;    // maxvr=25
  a[:m+4]=a[:m+1]+(sy/sran)*sqrt(0.90/0.10)*random;    // maxvr=10
  a[:m+5]=a[:m+1]+(sy/sran)*sqrt(0.95/0.05)*random;    // maxvr= 5

  (dates, times) = gen_date(n, 970101, 22, 72);

  // gen_date is a library func which generates 2 vectors of length n
  // The first vector includes dates from 970101 with 22 dates
  // per month.  The vector includes 72 time stamps (1 to 72) for
  // each date.  The dates and times are artificial and are included
  // to simulate features that require dates and times in FKR.

  a[:m+6]=times;
  a[:m+7]=dates;
  datecol = m + 7;
  boutput("chapter5.pri", datecol, a);  // creates a binary output file
  printf("chapter5.pri created");
}
main();
```

Figure 5.1 A TIMES program used to generate an artificial data file.[1]

signal in which the values are approximately half positive and half negative, the value of *FracSS* is meaningful: a value "significantly" above 0.5 is an indication that a model has predictive power. (Appendix E presents a statistical test for the significance of *FracSS*.)

Evaluation of the parameters and options discussed in Chapter 4 require software that implements these features. The FKR (Fast Kernel Regression) program[2] includes all these features, and a trial version can be downloaded from the Internet *(www.insightware.com)*. This program was used to generate the results shown in the following sections.

5.2 SEARCHING PARAMETERS

The parameters that control the range and depth of the search include *ncp* (the number of candidate predictors), *dmax* (the maximum dimensionality of all spaces considered) and *num_survivors(d)* (the maximum number of survivors from the search at dimensionality *d*). The total time of the analysis is approximately proportional to the total number of spaces considered. The "success" of the search is also dependent on the number of spaces, but the relationship is complex. The more spaces considered, the greater the probability of finding useful models. However, there is a point of diminishing returns. The value of *dmax* must be set high enough to find the good higher dimensional spaces if there are any. The values of *num_survivors(d)* must be high enough so that the good spaces (if there are any) at dimensionality *d* + 1 are not missed. But setting these parameters too high can lead to unacceptably high computer time for the analysis.

The magnitude of the problem is discussed in Section 3.6 and is illustrated using Table 3.10. In that table, the total number of spaces is shown as functions of *ncp* and *dmax*. Not included in the table is the total time required as functions of *ncp* and *dmax*. One would have to know T_{avg} (the average compute time required per space). For a given objective of T_{total} (total compute time), as T_{avg} increases the total number of spaces considered must decrease proportionally. The choice of the searching parameters becomes important only when the objective T_{total} becomes difficult to achieve. For small values of *ncp* the choice of searching parameters is irrelevant. Even if every possible space is examined, the value of T_{total} is still small. For example, for the data set created using the TIMES program shown in Figure 5.1, using the first 10 columns as candidate predictors (*ncp* = 10), and looking at all spaces from *d* = 1 to *d* = 10 (i.e., *dmax* = 10), the total number of spaces is 1274. This is a relatively small number of spaces and should not present a problem regarding total compute time. (This value can be verified using Eq. 3.42.)

The choice of a value for *dmax* is limited by *nlrn* (the number of learning data points) used to create the *p*-tree. The number of leaf cells in the tree is a function of *h* (the treeheight). The number of leaf cells is 2^h, and *ncell* (the average number of learning points per cell) is thus $nlrn/2^h$. If we choose a desired value of *ncell,* we can then determine the appropriate value of *h:*

$$h = \log2 \, (nlrn \, / \, ncell) \qquad (5.5)$$

For example, if *nlrn* = 10000 and we choose a value of *ncell* = 20, then $h = \log2(500) = 8.97$. Clearly, *h* must be an integer so we can choose a value of $h = 8$, which would make *ncell* = 39.1 or $h = 9$, which would make *ncell* = 19.5. (In the figures below, *ncell* is called BUCKETSIZE). Once *h* has been selected, then the upper limit for *dmax* is just equal to *h*. It is illogical to choose a value of *dmax* > *h* because the space is only partitioned *h* times. If *d* (the dimensionality of a space) exceeds *h,* then *d* − *h* of the dimensions will not be partitioned.

The choice of *dmax* does not automatically have to be set at *h*. The analyst might prefer to limit the complexity of the model by limiting *dmax*. An alternative approach to limiting model complexity is to use the δ concept introduced in Section 3.6. The δ parameter is the minimum improvement in *MC* (the modeling criterion) that a space must exhibit to be considered for survival. All surviving spaces examined at dimensionality *d* (if *d* > 1) must have a value of *MC* at least δ greater than its subspaces of dimensionality *d* − 1. If no spaces survive at dimensionality *d*, then the analysis is terminated even if *d* < *dmax*.

An example of the entire process is shown in Figure 5.2. This report is the output of the FKR analysis of the data generated using Figure 5.1. Note that the process is applied to Column 14, which is 10 percent signal and 90 percent noise. There are a number of points to note in Figure 5.2. The most important is that the best model located consists of the three candidate predictors that were used to create the true signal column (i.e., *X*2, *X*5, and *X*9). Furthermore, the Variance Reduction noted using this particular space is 5.3 percent, which is 53 percent of the theoretical value of 10 percent. This model exhibited a *FracSS* (Fraction Same Sign) of 0.562, which means that for over 56 percent of the 5000 test points, the sign of the predicted value of *Y* was the same as the actual value. The *F Statistic* value of 3.09 is considerably greater than the 2σ limit of 1.13 listed in the report and can therefore be considered as highly significant.

```
Output report:   FKR Analysis Wed Sep 16 15:33:56 1998
                 Version 3.07  Sept 3, 1998
Parameters used in analysis    PARAMETER FILE: chapter5.fpa
                               INPUT DATA FILE: chapter5.pri
DMIN    - min number of dimensions              :      1
DMAX    - max number of dimensions              :      5
NCP     - number of candidate predictors        :     10
NCOL    - number of columns of data             :     17
NREC    - total number of records               :  15000
LTYPE   - Learning set type (G, M or S)         :      S
          G is growing, M is moving and S is static
NUMCELL- Number of cells searched for NUMNN      :      1
NUMNN   - Irrelevant because FAST mode specified
STARTVAR - starting variable                    :      1
DELTA   - Minimum incremental change      :      1.00
NODATA - For Ycalc (columns of YCF file):     -999.00
Kernel Regression parameters:
MOD_CRITERION- 1=VarRed, 2=CorCoeff, 3=CC_Origin:      1
FIT_ORDER - 0 is average,  1 & 2 are surfaces  :      0
     FAST option used: Ycalc is cell average
NUMK - Number of smoothing parameters           :      1
K(1) - Smoothing parameter 1                    :   1.00
YCOL - Column of dependent variable             :     14
Tree parameters:
BUCKETSIZE - design number per cell (computed)  :     19
TREEHEIGHT - tree parm                          :      9
Computed number of leaf cells                   :    512
Computed 2 Sigma limit for F Stat               :   1.13
Data set parameters:
NLRN    - number of learning records           :  10000
NTST    - number of test records               :   5000
NEVL    - number of evaluation records         :      0
STARTLRN - starting learning record            :      1
STARTTST - starting test record                :  10001


Ordered results for 1 dimensional models:
    Total number of combinations:      10
    Number tested           :      10
    survivenum( 1)          :      10
    Number of survivors     :      10
Var Red:   -0.323   FracSS: 0.538   F:  1.64   X:   9
Var Red:   -1.074   FracSS: 0.538   F:  1.77   X:   5
Var Red:   -2.212   FracSS: 0.526   F:  1.58   X:   2
Var Red:   -4.195   FracSS: 0.504   F:  0.93   X:   3
Var Red:   -5.053   FracSS: 0.500   F:  0.99   X:   6
Var Red:   -5.139   FracSS: 0.495   F:  0.97   X:  10
Var Red:   -5.205   FracSS: 0.498   F:  0.94   X:   8
Var Red:   -5.520   FracSS: 0.501   F:  1.09   X:   1
Var Red:   -5.989   FracSS: 0.490   F:  1.04   X:   7
Var Red:   -6.001   FracSS: 0.494   F:  1.07   X:   4
```

Figure 5.2   FKR report for data generated using Figure 5.1 (part 1).[2]

For this particular example, the choice of the searching parameters was effective. The values of *survivenum(d)* were 10 for $d = 1$ and 5 for all other values of d. By choosing *survivenum*(1) = *ncp* (the number of candidate predictors), we ensure that all two-dimensional spaces are examined. For values of $d > 1$, the value of 5 restricted the search, but from the

```
Ordered results for 2 dimensional models:
   Total number of combinations:      45
   Number tested              :       45
   survivenum( 2)             :        5
   Number of survivors        :        5
Var Red:   2.489    FracSS: 0.547    F:  2.41    X:  5    9
Var Red:   2.296    FracSS: 0.549    F:  2.34    X:  2    5
Var Red:   1.267    FracSS: 0.549    F:  2.36    X:  2    9
Var Red:  -3.969    FracSS: 0.512    F:  0.95    X:  1    7
Var Red:  -3.997    FracSS: 0.507    F:  0.97    X:  6    10

Ordered results for 3 dimensional models:
   Total number of combinations:     120
   Number tested              :       38
   survivenum( 3)             :        5
   Number of survivors        :        5
Var Red:   5.298    FracSS: 0.562    F:  3.09    X:  2    5    9
Var Red:  -0.447    FracSS: 0.536    F:  1.72    X:  5    6    10
Var Red:  -0.598    FracSS: 0.537    F:  1.65    X:  6    9    10
Var Red:  -0.845    FracSS: 0.530    F:  1.65    X:  1    5    7
Var Red:  -1.898    FracSS: 0.527    F:  1.59    X:  2    6    10

Ordered results for 4 dimensional models:
   Total number of combinations:     210
   Number tested              :       32
   survivenum( 4)             :        5
   Number of survivors        :        4
Var Red:   2.318    FracSS: 0.540    F:  2.12    X:  5    6    9    10
Var Red:   2.137    FracSS: 0.556    F:  2.31    X:  1    5    7    9
Var Red:   1.721    FracSS: 0.556    F:  2.18    X:  2    5    6    10
Var Red:   0.287    FracSS: 0.544    F:  2.34    X:  1    2    5    7

Ordered results for 5 dimensional models:
   Total number of combinations:     252
   Number tested              :       22
   survivenum( 5)             :        5
   Number of survivors        :        3
Var Red:   3.964    FracSS: 0.564    F:  2.82    X:  2    5    6    9    10
Var Red:   1.601    FracSS: 0.548    F:  2.34    X:  1    2    4    5    7
Var Red:   1.387    FracSS: 0.546    F:  2.17    X:  1    2    5    7    8

Best Model Report Fold 1 (Test Set Data):
Model: 1 Var Red:   5.298    FracSS: 0.562    F: 3.09    X: 2  5  9
Model: 2 Var Red:   3.964    FracSS: 0.564    F: 2.82    X: 2  5  6  9  10
Model: 3 Var Red:   2.489    FracSS: 0.547    F: 2.41    X: 5  9
```

Figure 5.2 FKR report for data generated using Figure 5.1 (part 2).

results we see that for this example this value allowed the program to find the true model. In fact, for this example, using *survivenum(d)* = 1 for all values of d would have led to the same model: only $X9$ would have survived the d = 1 search, then the $X5$ and $X9$ space would have survived d = 2, and finally $X2$, $X5$, and $X9$ would have been located in the 3D search. The choice of δ = 1 eliminated some spaces, and as a result, less than five survivors were noted for the 4D and 5D searches.

The searching strategy illustrated in this example is not always so successful. Some models show no predictive power for subspaces of the model. A minor modification to the TIMES program in Figure 5.1 creates such a model. In Figure 5.1 the pure signal in that program is the sum of three other terms:

```
a[:m+1] = y1+y2+y3;
// The pure signal is a 3D non-linear function
```

The y1 term is a function of the second column:

```
y1=exp (- ((a[:2] + 0.5) ^ 2))
    - exp (- ((a[:2] - 0.5) ^ 2));
```

Similarly, the y2 and y3 terms are functions of the fifth and ninth columns. If we replace this model by the product of the three terms and not the sum:

```
a[:m+1] = y1*y2*y3;
// The pure signal is a 3D non-linear function
```

we get completely different results as shown in Figure 5.3. The three subspaces of y are the 2D subspaces $(X2, X5)$, $(X2, X9)$, and $(X5, X9)$, and these subspaces have no predictive power. The expected value of y for any point in any of these subspaces is zero. Using the same values of *survivenum(d)*, we find that the true signal space $(X2, X5, X9)$ is not located. In this example, the δ parameter is set to a large negative number to eliminate the effect of this parameter on the search.

Even though the one-dimensional spaces $X9$ and $X5$ were the best spaces at this level, none of the 2D subspaces of the true signal space $(X2, X5, X9)$ survived the 2D search. As a result, the search failed to locate $(X2, X5, X9)$ at the 3D level. However, the 4D search did find $(X2, X5, X8, X9)$, which includes the true signal space. All five survivors of the 5D search included $(X2, X5, X9)$. Further testing of this example showed that a value of *survivenum(2)* = 12 was required to get one of the three subspaces to survive the 2D search. The Variance Reduction for $(X2, X5, X9)$ is 4.87, the *FracSS* is 0.557, and the *F Statistic* is 2.66.

The conclusion from the first example should be that some models can be easily located through judicious use of the vari-

```
Output report:  FKR Analysis Wed Sep 16 17:05:16 1998
                 Version 3.07  Sept 3, 1998
 Parameters used in analysis    PARAMETER FILE: chapter5.fpa
                                INPUT DATA FILE: chapter5.pri
 DMIN   - min number of dimensions           :     1
 DMAX   - max number of dimensions           :     5
 NCP    - number of candidate predictors     :    10
 NCOL   - number of columns of data          :    17
 NREC   - total number of records            : 15000
 LTYPE  - Learning set type (G, M or S)      :     S
          G is growing, M is moving and S is static
 NUMCELL- Number of cells searched for NUMNN  :     1
 NUMNN  - Irrelevant because FAST mode specified
 STARTVAR - starting variable                :     1
 DELTA  - Minimum incremental change    :   -100.00
 NODATA - For Ycalc (columns of YCF file):   -999.00
 Kernel Regression parameters:
 MOD_CRITERION- 1=VarRed, 2=CorCoeff, 3=CC_Origin:   1
 FIT_ORDER - 0 is average, 1 & 2 are surfaces  :   0
    FAST option used: Ycalc is cell average
 NUMK - Number of smoothing parameters       :     1
 K(1) - Smoothing parameter 1                : 1.00
 YCOL - Column of dependent variable         :    14
 Tree parameters:
 BUCKETSIZE - design number per cell (computed) :  19
 TREEHEIGHT - tree parm                      :     9
 Computed number of cells                    :  1023
 Computed number of leaf cells               :   512
 Computed avg bucket size                    :  19.5
 Computed 2 Sigma limit for F Stat           :  1.13
 Data set parameters:
 NLRN   - number of learning records         : 10000
 NTST   - number of test records             :  5000
 NEVL   - number of evaluation records       :     0
 STARTLRN - starting learning record         :     1
 STARTTST - starting test record             : 10001
 Ordered results for 1 dimensional models:
    Total number of combinations:      10
    Number tested            :         10
    survivenum( 1)           :         10
    Number of survivors      :         10
 Var Red:  -3.621   FracSS: 0.511   F:  0.88   X:   9
 Var Red:  -4.302   FracSS: 0.508   F:  0.98   X:   5
 Var Red:  -4.821   FracSS: 0.501   F:  0.96   X:   3
 Var Red:  -5.254   FracSS: 0.506   F:  1.04   X:   1
 Var Red:  -5.284   FracSS: 0.503   F:  0.98   X:  10
 Var Red:  -5.315   FracSS: 0.495   F:  0.90   X:   2
 Var Red:  -5.542   FracSS: 0.493   F:  0.94   X:   8
 Var Red:  -5.683   FracSS: 0.496   F:  1.01   X:   7
 Var Red:  -5.701   FracSS: 0.496   F:  1.05   X:   6
 Var Red:  -5.949   FracSS: 0.490   F:  1.06   X:   4
```

Figure 5.3 FKR results for y = y1*y2*y3, 10% signal, 90% noise (part 1).

ous parameters employed. However, the second example illustrates the point that some models cannot be found so easily. This is particularly true of models in which the subspaces of the models have little predictive power. For such cases, a much more exhaustive search is required. The problem faced by the analyst is that one never knows ahead of time if a model does exist, and if so, what is the nature of the model!

```
Ordered results for 2 dimensional models:
   Total number of combinations:      45
   Number tested           :          45
   survivenum( 2)          :           5
   Number of survivors      :          5
Var Red:  -3.886   FracSS: 0.503   F:  0.96   X:   9  10
Var Red:  -3.934   FracSS: 0.503   F:  0.95   X:   1   7
Var Red:  -3.946   FracSS: 0.509   F:  1.04   X:   1   5
Var Red:  -4.034   FracSS: 0.500   F:  0.82   X:   3   9
Var Red:  -4.051   FracSS: 0.507   F:  0.96   X:   2   8


Ordered results for 3 dimensional models:
   Total number of combinations:     120
   Number tested           :          38
   survivenum( 3)          :           5
   Number of survivors      :          5
Var Red:  -3.626   FracSS: 0.507   F:  1.04   X:   1   7  10
Var Red:  -3.706   FracSS: 0.507   F:  0.96   X:   2   5   8
Var Red:  -3.968   FracSS: 0.500   F:  0.81   X:   2   8  10
Var Red:  -4.139   FracSS: 0.509   F:  1.00   X:   1   2   8
Var Red:  -4.208   FracSS: 0.512   F:  0.92   X:   3   6   9


Ordered results for 4 dimensional models:
   Total number of combinations:     210
   Number tested           :          32
   survivenum( 4)          :           5
   Number of survivors      :          5
Var Red:   0.206   FracSS: 0.554   F:  2.26   X:   2   5   8   9
Var Red:  -3.525   FracSS: 0.500   F:  0.91   X:   3   6   9  10
Var Red:  -3.857   FracSS: 0.516   F:  0.93   X:   3   4   6   9
Var Red:  -3.988   FracSS: 0.507   F:  0.92   X:   1   2   6   8
Var Red:  -4.159   FracSS: 0.521   F:  0.99   X:   1   3   7  10


Ordered results for 5 dimensional models:
   Total number of combinations:     252
   Number tested           :          29
   survivenum( 5)          :           5
   Number of survivors      :          5
Var Red:   0.420   FracSS: 0.547   F:  1.98   X:   2   3   5   8   9
Var Red:   0.145   FracSS: 0.556   F:  2.11   X:   2   5   7   8   9
Var Red:   0.050   FracSS: 0.551   F:  1.94   X:   2   5   8   9  10
Var Red:  -0.050   FracSS: 0.550   F:  2.05   X:   2   4   5   8   9
Var Red:  -0.069   FracSS: 0.561   F:  1.93   X:   1   2   5   8   9


Best Model Report Fold 1 (Test Set Data):
Model: 1 Var Red: 0.420   FracSS: 0.547   F: 1.98   X:  2   3   5   8   9
Model: 2 Var Red: 0.206   FracSS: 0.554   F: 2.26   X:  2   5   8   9
Model: 3 Var Red: 0.145   FracSS: 0.556   F: 2.11   X:  2   5   7   8   9
```

Figure 5.3   FKR results: y = y1*y2*y3, 10% signal, 90% noise (part 2).

5.3 THE EFFECT OF TREE HEIGHT

The treeheight parameter h has an important effect on compute time. In Section 4.8 the calculational complexity of kernel regression analyses was discussed. The average time per space T_{avg} consists of the preparation time and the run time as shown in Eq. (4.5). In the discussion following this equation, it was explained that an important factor in the T_{prep} term is the time required to sort the data, and Eq. (4.2) is a maximum value for this sort time. Values of the relative maximum sort time for various combinations of h and n (number of learning points in the tree) are included in Table 5.1.

In this table, all values are relative to $n = 2^8 = 256$ and $h = 5$. For example, the upper limit for the sort time for $n = 2^{16} = 65536$ and $h = 10$ is 981 times greater than the upper limit for the sorts required to create a tree of $h = 5$ using 256 points. Some entries in the table are listed as n.a. (i.e., not applicable). These entries are for combinations in which the number of leaves on the tree (i.e., 2^h) is greater than the number of points. The number of points per leaf must be at least one. One of the most interesting observations that one can note from Table 5.1 is the relatively small effect of h for a given value of n. For example, for $n = 2^{16} = 65536$ we see that the maximum sort time increases from 597 for $h = 5$ to 1109 for $h = 13$. This increase is less than a factor of 2, even though an additional eight levels of sorting are required!

The direct effect of treeheight on running time T_{run} is usually not significant. To locate the leaf in which a test point resides requires a series of h *if* statements, but the time required to do this is usually negligible. However, many other design implications are associated with the choice of h which do

TABLE 5.1 Relative Maximum Sort Times for Combinations of n and h

n	h=5	h=6	h=7	h=8	h=9	h=10	h=11	h=12	h=13
2^8	1.00	1.10	1.17	1.20	n.a.	n.a.	n.a.	n.a.	n.a.
2^9	2.33	2.60	2.80	2.93	3.00	n.a.	n.a.	n.a.	n.a.
2^{10}	5.33	6.00	6.53	6.93	7.20	7.33	n.a.	n.a.	n.a.
2^{11}	12.0	13.6	14.9	16.0	16.8	17.3	17.6	n.a.	n.a.
2^{12}	26.7	30.4	33.6	36.2	38.4	40.0	41.1	41.6	n.a.
2^{13}	58.7	67.2	74.7	81.1	86.4	90.7	93.9	96.0	97.1
2^{14}	128	147	164	179	192	203	211	218	222
2^{15}	277	320	358	393	422	448	469	486	499
2^{16}	597	691	777	853	922	981	1033	1075	1109
2^{17}	1280	1485	1673	1843	1997	2133	2253	2355	2441
2^{18}	2731	3174	3584	3959	4301	4608	4881	5120	5325
2^{19}	5803	6758	7646	8465	9216	9899	10513	11059	11537
2^{20}	12288	14336	16247	18022	19661	21163	22528	23757	24849

impact T_{run}. If we hold all other parameters the same, the effect of h on the results is quite significant.

To illustrate the effect of h on the results, consider Table 5.2 in which values of VR are listed as a function of treeheight. Note that this table was created using the Y values of the pure signal created in the program of Figure 5.1. For each of the three KR algorithms tested, the value of VR approaches 100 as h increases. All analyses in this section use equal weighting of the learning points in the test cells.

The limiting value of h is a function of n (the number of learning points), and for this example the value was 10,000. For the *Order 0* Algorithm, the average number of points per leaf cell must be at least equal to one, and we therefore see that the limiting value of h is 13. (For $h = 14$ the average number of points per leaf cell is $10000/2^{14} = 0.61$.) For the *Orders 1* and *2* algorithms there must be at least one more point than coefficients required to describe the computed surface. (In the table the message *numnn < min* indicates that this condition is not satisfied.) The table was generated using only the points in the test cells (i.e., the leaf cells in which each test point fell). Since the model is three dimensional, the number of coefficients required to define a plane (*Order 1*) in three dimensions is four and thus a minimum of five points per cell is required. The number of coefficients required to define a second-order multinomial (*Order 2*) in three dimensions is 10, and thus a minimum of 11 points per cell is required. Thus for *Order 1* the limiting value of h is 10, and for *Order 2* it is 9.

Table 5.2 shows that VR increases toward a value of 100 (i.e., a perfect model) as h approaches its limiting value. Furthermore, for a given value of h, VR increases as the order of the algorithm increases. The results for small values of h are explainable. The particular model generated in Figure 5.1 is a three-dimensional model with eight distinct regions. We can examine the equations used to generate the model:

```
y1=exp (- ((a[:2] + 0.5) ^2)) - exp (- ((a[:2] - 0.5) ^ 2));
y2=exp (- ((a[:5] + 0.5) ^2)) - exp (- ((a[:5] - 0.5) ^ 2));
y3=exp (- ((a[:9] + 0.5) ^2)) - exp (- ((a[:9] - 0.5) ^ 2));
a[:m+1] = y1+y2+y3;
// The pure signal is a 3D non-linear function
```

TABLE 5.2 Values of Variance Reduction for Combinations of Treeheight h and the Order of the Kernel Regression Algorithm Using Pure Signal Data

h	Leaf Cells	Avg/cell	VR (Ord 0)	VR (Ord 1)	VR (Ord 2)
13	8192	1.22	93.97	*numnn<min*	*numnn<min*
12	4096	2.44	93.70	*numnn<min*	*numnn<min*
11	2048	4.88	91.76	*numnn<min*	*numnn<min*
10	1024	9.77	89.72	98.68	*numnn<min*
9	512	19.53	84.92	98.04	99.42
8	256	39.06	71.31	96.02	98.86
7	138	78.13	61.15	92.73	98.30
6	64	156.25	53.37	89.22	97.26
5	32	312.50	53.37	76.70	86.13
4	16	625.00	52.89	64.85	78.61
3	8	1250.00	52.40	52.85	73.58
2	4	2500.00	−0.02	30.50	30.63
1	2	5000.00	−0.03	−0.03	18.02
0	1	10000.00	−0.00	−0.05	−0.02

Note that y1 is positive for values of column 2 < 0 and negative for values of column 2 > 0. The maximum value of y1 is close to the value obtained at a2 = −0.5, which is $1-e^{-1} = 0.632$. The minimum value is close to the value obtained at a2 = 0.5: −0.632. Similarly, y2 and y3 are also positive for negative values of columns 5 and 9. A three-dimensional picture of the pure signal would show four maxima and four minima. Using the *Order 0* Algorithm with unit weighting of all the points in a cell, we find that the cell average value is the predicted value for each test point. To model this signal using the *Order 0* Algorithm, at least eight leaf cells are required, and therefore we see that for $h <= 2$, no Variance Reduction is obtained. For *Order 1,* we can still get VR for $h = 2$. The slope of the plane within each cell compensates for the fact that only four cells are used to describe a surface with eight different regions. However, when the number of cells is reduced to two, then the *Order 1* Algorithm fails. Similarly, the *Order 2* Algorithm achieves significant VR with only two cells but fails when the number of cells is reduced to one. For this particular eight-region signal, we would require an *Order 3* Algorithm to achieve significant Variance Reduction using only one leaf cell.

One might get the mistaken impression from Table 5.2 that the correct strategy is to use the largest possible value of h and the highest order algorithm available. When the pure signal is corrupted with noise, the situation becomes more complex. The effect of h on VR with low signal-to-noise data is shown in Table 5.3. The results in this table were obtained using data generated with 10 percent signal and 90 percent noise.

The results in Table 5.3 are quite different from the results in Table 5.2. The first point to note is that the magnitudes of the values of VR are much less. Since the data is only 10 percent signal and the remainder is just noise, we expect the values of VR to be limited to 10. Even though the curves are not smooth owing to the statistical nature of the data, we can observe a general trend. For all three algorithms, we note that as h decreases from its maximum value, VR first increases and then decreases. In other words, for each algorithm there is an optimum value of h. For noisy problems, the number of learning points (i.e., *numnn*) used to make the predictions must be great enough to compensate for the noise in each individual learning point. This requirement

TABLE 5.3 Values of Variance Reduction for Combinations of Treeheight *h* and the Order of the KR Algorithm Using Data with 10% Signal and 90% Noise

h	Leaf Cells	Avg/cell	VR (Ord 0)	VR (Ord 1)	VR (Ord 2)
13	8192	1.22	−59.74	*numnn*<min	*numnn*<min
12	4096	2.44	−22.42	*numnn*<min	*numnn*<min
11	2048	4.88	−7.29	*numnn*<min	*numnn*<min
10	1024	9.77	−0.03	−50.67	*numnn*<min
9	512	19.53	4.87	−10.37	−167.71
8	256	39.06	4.10	−0.01	−29.45
7	138	78.13	4.66	5.02	−4.09
6	64	156.25	4.51	7.11	4.12
5	32	312.50	4.98	7.07	5.71
4	16	625.00	5.37	6.13	6.77
3	8	1250.00	5.34	5.24	7.34
2	4	2500.00	0.07	2.44	2.14
1	2	5000.00	0.02	−0.02	1.01
0	1	10000.00	−0.01	0.00	−0.10

favors smaller values of h. On the other hand, the larger the value of h, the smaller the region within each leaf cell and the lesser the variation of the signal within the cells.

One of the most encouraging conclusions from the results in Table 5.3 is that there is a fairly wide range of values of h in which VR is acceptable. For the *Order 0* results, the values of VR in the range $3 <= h <= 9$ are quite significant. They range from 41 percent to about 54 percent of the maximum theoretical VR for this data. For *Order 1* the values of VR range from 50 to 71 percent for $3 <= h <= 7$. For *Order 2* the values of VR range from 41 to 73 percent for $3 <= h <= 6$.

5.4 NUMBER OF NEAREST NEIGHBORS

The parameter *numnn* (number of nearest neighbors) affects both compute time and the quality of predictions. Using the data structure described in Chapter 4, we find that each test data point is associated with a leaf in the tree created from the learning data points. (Test points falling outside the data range of the tree are not used.) If the user is willing to limit the set of learning points used for making predictions to all the points in the associated leaf cells, very rapid predictions can be made for the test points. There is no need to initiate a search for the *numnn* points. One must only be able to access the points within each leaf cell. We call this mode of operation *fast mode*. By initiating the *fast mode* of operation, the quality of the resulting predictions is less than what one can obtain using a better choice of points. In this section the effect of specifying *numnn* is examined using the artificial data set generated by the program shown in Figure 5.1.

An important parameter defining the search for the *numnn* points is *numcells* (the number of cells in which the search is performed). If *numcells* is one, then only the single leaf cell associated with each test point is searched. If *numcells* is two, then the leaf cell plus the nearest adjacent cell (nearest with respect to each test point) are searched. Increasing the value of *numcells* to three increases the number of adjacent cells searched to two. In the FKR Program[2] used to obtain the results discussed in this chapter, if *numcells* is a large number, then all adjacent cells are included in the search. Cells that are not adjacent to the test point cell are never included in the search. The number of adjacent cells may vary from cell to cell. In Section 4.2 the concept of an *adjacency matrix* was introduced. If *numcells* is greater than one, an adjacency matrix must be generated. For a particular space, generation of this matrix must only be performed as part of the preparation phase and therefore only contributes to the preparation time T_{prep}. The matrix is then used for all test points. The time to generate the matrix T_{adj} is proportional to 2^{2h} (see Eq. 4.7), so for large values of h the term T_{adj} can become the dominant factor in determining the value of T_{prep}.

The parameters *numnn* and *numcells* have an important effect on run time T_{run}. From Eq. (4.11) we see that the constant C_{run} (the average run time per test point) includes the term C_{search} (the average search time for the *numnn* points if a search is initiated). If C_{search} is nonzero, this search time can be the dominating term in setting the value of T_{run}. Another alternative is available. One can use only the test leaf cell, but the value of *numnn* can be specified as less than the number of points in the cell. If this alternative is chosen, then there is no need to generate the adjacency matrix. However, T_{run} is still dominated by the time required for the searches. This alternative is usually associated with a small value of h and thus a large number of learning points in each leaf cell.

To understand the interplay of these parameters, first consider the case of the pure signal generated by the program in Figure 5.1. In Table 5.4 values of VR (Variance Reduction) are listed for various combinations of *numnn* and *numcells*. All results in this table are generated using the *Order 0* Algorithm with unit weighting (i.e., the prediction for each test point is the average value of the *numnn* learning point values of Y). It should be emphasized, however, that qualitatively similar results could be obtained if the *Order 1* and *2* algorithms were used. A tree of $h = 6$ was selected. Since there were 10,000 learning points, the average number of points per leaf cell is $10000/2^6 = 156.25$. All results were obtained using the three-dimensional space composed of $X2$, $X5$, and $X9$. The tree created for this space has 64 leaf cells. The number of adjacent cells for each of these leaf cells ranges from 4 to 17, with an average value of 8.78. Table 5.4 includes results that show the effect of specifying *numnn* and *numcells*. The highest value of VR for each column is shown in boldface type.

On can come to several obvious conclusions from the results in this table. The most obvious one is that for a pure signal, the greater the number of cells searched for the *numnn* nearest neighbors, the greater the resulting value of VR. However, the change in VR per change in *numcells* decreases with increasing *numcells*. We can also conclude that the optimum value of *numnn* is a small number. For this particular set of data, the optimum value of *numnn* varied from 4 for *numcells* = 1 to a value around 10 if *numcells* is large enough to include all adjacent cells for all 64 leaf cells. These results have a clear physical explanation. The greater the value of *numcells,* the better the selection of the *numnn* learning points. The small value of the optimum is due to the nonlinear shape of the surface of the pure signal. If *numnn* is large, then points that are relatively far from the test points are used to determine the average values (i.e., the predictions). In other words, for a pure signal it is better to use a few very close points rather than a lot of points, including many that are relatively far from the test points.

TABLE 5.4 Values of VR for Combinations of *numnn* and *numcells* for Pure Signal

numnn	numcells=1	numcells=2	numcells=5	numcells=10	All adj cells
1	97.07	97.12	97.28	97.41	97.43
2	97.79	97.83	98.02	98.22	98.24
3	97.96	98.02	98.24	98.47	98.52
4	**98.01**	**98.08**	98.33	98.63	98.70
5	97.95	98.07	**98.34**	98.71	98.78
10	97.29	97.59	98.10	**98.74**	**98.88**
20	94.97	95.92	97.09	98.42	98.68
40	88.71	91.28	94.48	97.35	97.85
75	76.02	81.62	89.19	95.17	96.03
150	54.18	60.82	77.53	90.01	91.24

The situation changes dramatically for the noisy types of problems encountered when trying to model financial markets. Table 5.5 is similar to Table 5.4 but is based on the data column that is 10 percent signal and 90 percent noise. The maximum VR that one could hope for in modeling this data is VR=10. In this table, for a given value of *numnn* the results tend to improve with increasing *numcells*. However, this does not always happen owing to the statistical nature of the noise. The optimum values of *numnn* are in the 50 to 100 range, which is much greater than the optimums observed in Table 5.4 for the pure signal. Once again, the physical interpretation of these results is clear. The value of *numnn* required to maximize VR increases with increasing noise because a larger number of points are required to obtain meaningful averages.

The results in Table 5.5 should be compared to the value of 4.51 noted in Table 5.3 for *h* = 6 and for *Order 0*. The result for *numnn* = 150 and *numcells* = 1 (i.e., 4.60) is slightly better because the six or seven points in the cell furthest away from each test point are not used. Much better values of VR can be obtained by choosing an optimum value of *numnn* and using the adjoining cells in the search for the nearest neighbors. What this table does not show is the price that must be paid for this more extensive search. For this particular example the cost is consid-

TABLE 5.5 Values of VR for Combinations of *numnn* and *numcells* for Data with 10% Signal and 90% Noise.

numnn	numcells=1	numcells=2	numcells=5	numcells=10	All adj cells
5	−9.21	−9.20	−7.91	−8.67	−8.84
10	1.08	1.39	1.86	1.65	1.35
15	4.29	4.44	5.03	4.75	4.80
20	5.80	5.79	6.16	6.11	6.37
30	6.96	7.09	7.47	7.62	7.77
50	**7.09**	**7.33**	**8.52**	8.99	9.11
75	6.75	7.23	8.43	**9.28**	**9.29**
100	6.13	6.75	8.12	9.13	9.21
125	5.35	6.12	7.89	9.20	9.17
150	4.60	5.31	7.57	9.15	9.14

Bold face signifies best value in column.

erable. Using the *fast mode* option (i.e., use all points in the test cells), the time to obtain the result VR = 4.51 was less than one second on a Pentium 100 computer. Using a value of *numnn* = 50, we see in Table 5.5 that the results are better and range from VR = 7.09 to 9.11. The time required to obtain these results was 5 seconds for *numcells* = 1, 6 seconds for *numcells* = 2, 12 seconds for *numcells* = 5, 19 seconds for *numcells* = 10, and 21 seconds if all adjacent cells are used. Thus for this particular data set a considerable computing cost is associated with exploiting the parameters *numnn* and *numcells*. If the number of spaces to be examined is small (i.e., the number of candidate predictors *ncp* is small), then this computing cost should not be an intolerable burden. However, if many spaces are to be examined, then a more cost-conscious strategy might be employed. One possibility is to use the *fast mode* option to hone in on the better spaces and then examine them in greater detail using the parameters *numnn* and *numcells*. Timing considerations associated with these parameters are considered in greater detail in Section 5.5.

5.5 PROCESSING TIME PER SPACE

The results in this section are based on timing studies using the data set generated by the program in Figure 5.1. It is well known that timing studies on computers are complicated. A review of the problems associated with timing studies is included in an interesting paper on the subject by Kernighan and Van Wyk.[3] They summarize the problems as follows: "The timing services provided by programs and operating systems are woefully inadequate. It is difficult to measure runtimes reliably and repeatably even for small, purely computational kernels, and it becomes significantly harder when a program does much I/O or graphics." To minimize these problems, the timer in FKR is initiated after data input is completed and output can be kept small. In addition, the program does not include a graphics module. Nevertheless, it should be recognized that the results are subject to variation when repeated. Results that require differences of two separate measurements are especially problematic. Meaningful results can, however, be obtained by making sure that each measurement takes enough time so that the results are significant.

The average time per space T_{avg} can be very fast if the *fast mode* of operation is used. This *fast mode* option avoids the need for a computer-intensive search for nearest neighbors. Once a space has been partitioned into cells, each test point is associated with a single cell (unless it falls outside the range of the learning points). Thus the cell becomes the "neighborhood," and the learning points within the cell become the nearest neighbors. Furthermore, the points in the cell are equally weighted so that a single surface is used for the entire space within the cell. As a result, the value of T_{avg} (see Eq. 4.5) is primarily the preparation time T_{prep}. The run time T_{run} is very small even for large values of *ntst* (number of test points). The value of T_{run} is proportional to *ntst* (Eq. 4.10), and the constant C_{run} (run time per test point) includes a number of components (see Eq. 4.11). If the *fast mode* option is used, the terms C_{search} and C_{weight} are zero and the terms C_{cell} and C_{solve} are small. The term C_{solve} only includes evaluation of *y*. There is no need to regenerate the matrix and solve the matrix equation. Using Eq. (3.7), we find that the known values of the coefficients A_j are used to compute *y*.

Results using the *fast mode* option and a Pentium 100 processor are included in Figures 5.4 to 5.6. These figures were extracted from the output report of the FKR program. (Interestingly, the runs were repeated using a Pentium II=400, and the speedup was a factor of over 6.) The relevant parameters were *h* = 8, *nlrn* = 10000, and *ntst* = 5000. The search was limited by

```
Timing Report:
  dim: 1  Num_spaces:   10   Time:    2   Time/space:  0.20
  dim: 2  Num_spaces:   45   Time:   31   Time/space:  0.69
  dim: 3  Num_spaces:   37   Time:   28   Time/space:  0.76
  dim: 4  Num_spaces:   29   Time:   24   Time/space:  0.83
  dim: 5  Num_spaces:   28   Time:   25   Time/space:  0.89
  dim: 6  Num_spaces:   15   Time:   13   Time/space:  0.87
  dim: 7  Num_spaces:   16   Time:   15   Time/space:  0.94
  dim: 8  Num_spaces:   11   Time:   11   Time/space:  1.00
  dim: 9  Num_spaces:    5   Time:    5   Time/space:  1.00
  dim:10  Num_spaces:    1   Time:    1   Time/space:  1.00
  Total:  Num_spaces:  197   Time:  155   Time/space:  0.79
```

Figure 5.4 Timing report from FKR output report: *order* = 0, *fast mode*.

```
Timing Report:
   dim: 1  Num_spaces:   10   Time:    3   Time/space:  0.30
   dim: 2  Num_spaces:   45   Time:   32   Time/space:  0.71
   dim: 3  Num_spaces:   36   Time:   29   Time/space:  0.81
   dim: 4  Num_spaces:   33   Time:   28   Time/space:  0.85
   dim: 5  Num_spaces:   25   Time:   23   Time/space:  0.92
   dim: 6  Num_spaces:   23   Time:   22   Time/space:  0.96
   dim: 7  Num_spaces:   18   Time:   19   Time/space:  1.06
   dim: 8  Num_spaces:   13   Time:   14   Time/space:  1.08
   dim: 9  Num_spaces:    6   Time:    7   Time/space:  1.17
   dim:10  Num_spaces:    1   Time:    1   Time/space:  1.00
   Total:  Num_spaces:  210   Time:  178   Time/space:  0.85
```

Figure 5.5   Timing report from FKR output report: *order* = 1, *fast mode.*

```
Timing Report:
   dim: 1  Num_spaces:   10   Time:    4   Time/space:  0.40
   dim: 2  Num_spaces:   45   Time:   34   Time/space:  0.76
   dim: 3  Num_spaces:   35   Time:   34   Time/space:  0.97
   dim: 4  Num_spaces:   32   Time:   38   Time/space:  1.19
   dim: 5  Num_spaces:   24   Time:   36   Time/space:  1.50
   dim: 6  Num_spaces:   19   Time:   39   Time/space:  2.05
   dim: 7  Num_spaces:   17   Time:   49   Time/space:  2.88
   Total:  Num_spaces:  182   Time:  234   Time/space:  1.29
```

Figure 5.6   Timing report from FKR output report: *order* = 2, *fast mode.*

setting *num survivors(d)* = 5 for all values of $d > 1$. Note that the effect of the *order* of the algorithm is small. Also, the effect of dimensionality d is also small for values of $d > 1$. For $d = 1$, only one sort is required regardless of the treeheight h, so T_{prep} is considerably smaller. Comparing the value of T_{avg} for $d = 2$ and $d = 10$, we see that the ratio T_{avg} $(d = 2)/T_{avg}$ $(d = 10)$ is about 0.7 for both *order* = 0 and *order* = 1. The results for *order* = 2 show that the maximum dimensionality possible for this example was $d = 7$. This limit is due to the minimum number of points required to compute the coefficients of the surface. For h = 8 and *nlrn* = 10000, the average number of points per cell *ppc* is 39.5. For *order* = 2, the minimum number of points for $d = 8$ exceeds this value. The number of terms in Eq. (3.7) increases rapidly as d increases, so C_{solve} increases proportionally.

In Section 5.4 the effects associated with the parameters *numnn* (number of nearest neighbors) and *numcells* (the number of cells in which the search is performed) were discussed. By setting *numcells* to a value greater than one and then using a value of *numnn* close to optimum, the observed VR (variance reduction) can be significantly increased. However, it was noted that if these parameters are used usually the required compute time increases considerably. The results in Figures 5.7 and 5.8 show the increased time required when a nearest-neighbor search is initiated.

Comparing the results in Figures 5.4 and 5.7, we can explain the difference by the added cost associated with a nearest neighbor search for the test data set. For $numcells = 1$ the dependence on d for $d > 1$ is small. Even for $d = 1$ the value of T_{avg} is over 50 percent of the value for $d = 4$. However, when $numcells$ is large enough so that each adjacent cell is searched for every test point, there is a dramatic increase in the value of T_{avg}, and the effect of dimensionality d becomes important. As d increases, the average number of adjacent cells increases rapidly, causing much greater values of C_{search} and therefore T_{avg}. Comparing the values of T_{avg} for $d = 4$ from Figures 5.4 and 5.8, we see an increase by a factor of about 15.

Equation (4.5) apportions the average time per space T_{avg} to two terms: running time T_{run} and preparation time T_{prep}. The T_{prep} term is the same regardless of whether or not the *fast mode* option is used. The terms contributing to T_{prep} are shown in eq. (4.6). In particular, the two important terms are *Sort_Time* and T_{adj} (the time to generate the adjacency matrix)

```
Timing Report:
    dim: 1   Num_spaces:    10    Time:     9    Time/space:  0.90
    dim: 2   Num_spaces:    45    Time:    65    Time/space:  1.44
    dim: 3   Num_spaces:    35    Time:    55    Time/space:  1.57
    dim: 4   Num_spaces:    32    Time:    54    Time/space:  1.69
    Total:   Num_spaces:   122    Time:   183    Time/space:  1.50
```

Figure 5.7 Timing report: *order = 0, numnn = 15, numcells = 1.*

```
Timing Report:
    dim: 1   Num_spaces:    10    Time:    19    Time/space:   1.90
    dim: 2   Num_spaces:    45    Time:   200    Time/space:   4.44
    dim: 3   Num_spaces:    37    Time:   302    Time/space:   8.16
    dim: 4   Num_spaces:    30    Time:   383    Time/space:  12.77
    Total:   Num_spaces:   122    Time:   904    Time/space:   7.41
```

Figure 5.8 Timing report: *order = 0, numnn = 15, numcells = all adjacent cells.*

if *numcells* is greater than one. We can measure the values of these terms using the data set introduced in Figure 5.1 and the FKR implementation of the data structure discussed in Chapter 4. The results are specific to this particular implementation and to the hardware used to obtain the results.

To measure T_{prep}, the easiest procedure is to make T_{run} as small as possible. Since T_{run} is proportional to the number of test points *ntst* (see Eq. 4.10), all one needs to do is set *ntst* to 2 (the smallest value permitted by the FKR program). Another problem associated with measuring T_{prep} is that the actual values in seconds are very small. The suggested procedure is to perform the measurement over a number of spaces. Using the data set generated by the program included in Figure 5.1, we obtain results for all 120 three-dimensional spaces using all 10 candidate predictors. (The number of combinations of 10 things taken three at a time is 10*9*8/6 = 120.) In other words, to determine the values of T_{prep}, the measured times are divided by 120. To measure the values of T_{adj}, one need only set *numcells* to two and *nlrn* (the number of learning points used to create the tree) to the minimum value for each value of treeheight *h*. (The minimum value is equal to the number of leaf cells on tree: i.e., 2^h.) To eliminate the effect of *Sort_Time* and T*book* (bookkeeping time), the measurement is repeated using *numcells* = *1* and the same value of *nlrn*. The difference is 120 times T_{adj}.

Let's turn our attention to *Sort_Time;* this term is a function of *nlrn* and *h* as seen in Eq. (4.2). Even though this equation is shown as an inequality, for higher dimensions (like 3D) it is close to an equality. We can define *nlrn_critical* as the value of *nlrn* at which *Sort_Time* is equal to T_{adj}. Values of T_{adj} and *nlrn_critical* are included in Table 5.6 as a function of *h*. The actual value of the constant *C* in Eq. (4.2) was determined to be $9.31*10^{-7}$, but this value is only applicable to the hardware configuration used to determine the results (a Pentium 100). Similarly, the values of T_{adj} are machine dependent. The results for *nlrn_critical* should be machine independent.

Measured results for T_{prep} are included in Table 5.7 for *numcells* = 1 and *numcells* > 1. The tremendous difference between the two columns for large values of *h* is due to the effect of T_{adj},

TABLE 5.6 Values of T_{adj} and *nlrn_critical* as a Function of *h*

h	T_{adj}	nlrn_critical
5	0.0012	41
6	0.0047	161
7	0.0189	480
8	0.0755	1440
9	0.3022	4440
10	1.209	14000
11	4.835	45120
12	19.34	148240
13	77.36	494850

which increases as 2^{2h} and therefore becomes increasingly dominant as *h* increases.

The run time T_{run} can be very small if the *fast mode* of operation is chosen. However, if we wish to use the full power of the software and perform a "nearest neighbor" search for learning points, then T_{run} can be large. The value of T_{run} is proportional to *ntst* (Eq. 4.10), and the constant C_{run} (run time per test point) includes a number of components (see Eq. 4.11). The discussion after Eq. (4.11) details the effects of various parameters on these components. To allow the user to make some time estimates, results from measurements using the data set generated from the program in Figure 5.1 are presented.

TABLE 5.7 Values of T_{prep} and as a Function of *h*: nlrn = 10000

h	T_{prep} (numcells = 1)	T_{prep} (numcells>1)
5	0.467	0.467
6	0.525	0.525
7	0.567	0.583
8	0.633	0.692
9	0.641	0.914
10	0.700	1.686
11	0.733	4.714
12	0.808	18.200
13	0.892	78.250

Table 5.8 includes values of run time per 1000 test points (i.e., $1000C_{run}$) for combinations of *numnn* and *h*. All values were determined using *numcells* = 1, *nlrn* = 10000, and the *Order 0* Algorithm. The results are from measurements made using a Pentium 100 and so should be scaled accordingly to estimate C_{run} for other hardware. Only three-dimensional spaces were used for this set of measurements. Some of the entries in the table are listed as n.a. (i.e., not applicable). When the value of *numnn* exceeds *ppc* (the average number of learning points per cell), then *numcells* = 1 is no longer applicable.

Table 5.9 includes values of run time per 1000 test points for combinations of *numcells* and *h*. All values were determined using *numnn* = 1, *nlrn* = 10000, and the *Order 0* Algorithm. In this table values of *h* were limited to 10. For values of *numcells* > 1 and *h* > 10 the preparation time T_{prep} is so dominant that the run time even for 5000 test points becomes insignificant and is difficult to measure with any reasonable level of accuracy.

The results in Tables 5.8 and 5.9 show a strong dependence of T_{run} on *h*. This behavior is due to *ppc* (points per cell). As *h* increases, *ppc* decreases dramatically (see Eq. 4.8). Thus as *h* increases, the number of learning points examined to find the nearest neighbors decreases. The time required to search for the nearest neighbors (i.e., C_{search}) is an important component of the run time unless the *fast mode* option is used. For the results shown in the tables, it can be seen that this effect is very important. Similarly, the results in Table 5.9 show that as *num-*

TABLE 5.8 Run Time per 1000 Test Points: *numcells* = 1, *order* = 0

h	*ppc*	*numnn=1*	*numnn=8*	*numnn=32*	*numnn=64*
5	312.50	0.825	0.907	1.167	1.587
6	156.25	0.455	0.517	0.695	0.935
7	78.13	0.237	0.281	0.397	0.507
8	39.06	0.133	0.165	0.239	n.a.
9	19.53	0.096	0.100	n.a.	n.a.
10	9.77	0.054	0.071	n.a.	n.a.
11	4.88	0.042	n.a.	n.a.	n.a.
12	2.44	0.040	n.a.	n.a.	n.a.
13	1.22	0.042	n.a.	n.a.	n.a.

TABLE 5.9 Run Time per 1000 Test Points: *numnn* = 1, *order* = 0

h	numcells=1	numcells=2	numcells=5	numcells=10	all adj cells
5	0.825	1.607	3.927	6.067	6.307
6	0.455	0.865	2.035	3.555	4.115
7	0.237	0.463	1.063	1.923	2.363
8	0.133	0.262	0.582	1.072	1.343
9	0.096	0.160	0.326	0.587	1.757
10	0.054	0.123	0.200	0.371	0.478

cells increases, the run time increases quite significantly. Once again this dependency can also be attributed to the fact that as *numcells* increases, the number of learning points examined also increases. Table 5.8 shows that the parameter *numnn* also affects run time but not as strongly as *h* or *numcells*. Varying *numnn* does not change the number of points that are examined; it only affects the time to manage a list of length *numnn*. The results show that this time does increase as *numnn* increases, but it has a smaller overall impact on run time as compared to *h* and *numcells*.

Table 5.10 includes timing results in which the *order* of the algorithm and the dimensionality *d* of the space are varied. These results show several interesting effects. For *Order 0* we see that the increase in T_{avg} going from $d = 3$ to $d = 5$ is small.

TABLE 5.10 T_{avg} for Combination of *order, d,* and *numnn* for 1000 Test Points

order	d	numnn=15	numnn=20	numnn=30	numnn=40
0	3	1.628	1.667	1.829	2.800
0	4	1.714	1.728	1.900	3.057
0	5	1.768	1.804	2.000	3.268
1	3	2.114	2.214	2.600	3.914
1	4	2.314	2.457	2.886	4.429
1	5	2.553	2.696	3.238	4.857
2	3	3.457	3.771	5.486	6.910
2	4	*n.a.*	5.800	7.467	10.800
2	5	*n.a.*	*n.a.*	12.500	16.333

However, for *Order 1* the change is larger and is quite apparent for *Order* 2. This effect is due to the method of solution. For each *order*, a matrix must be generated, and then a matrix equation must be solved. For *Order 0* the matrix is simply 1 by 1 and a single equation is solved. For *Order 1*, the matrix is $d + 1$ by $d + 1$, and for *Order 2* the matrix is *size* by *size* where *size* is $1 + d + d* (d + 1)/2$. For $d = 3$ *size* = 10, for $d = 4$ *size* = 15, and for $d = 5$ *size* = 21. The time required to solve a set of *size* simultaneous equations is O($size^3$), which explains the strong dependence of T_{avg} on d for *order 2*. Some of the entries in the table are listed as n.a. (i.e., not applicable). When the value of *numnn* is less than *size* + 1, the FKR program aborts processing. The value of *size* for *Order* 2 and $d = 4$ causes this problem for *numnn* = 15 and for $d = 5$ for *numnn* = 15 and 20.

The effect of *numnn* on T_{avg} is less dramatic but increases as *order* increases. This effect is explained by the time required to generate the matrix. The number of terms in the matrix increases as O($size^2$), so the time to generate the matrix increases as O($numnn* size^2$). Another explainable phenomenon is the large jump in T_{avg} in going from *numnn* = 30 to *numnn* = 40. The average number of points per cell (*ppc*) is only $10000/2^h = 39.06$, so *numcells* had to be increased to 2 to allow a search for 40 points. This increase in *numcells* from 1 to 2 causes a small increase in T_{prep} (see Table 5.7) but also double the number of points examined in the search for the *numnn* nearest neighbors.

5.6 DATA WEIGHTING

The concept of data weighting was introduced in Section 3.1. Equation (3.1) is the exponential kernel used to weight data spatially. Time weighting of data was introduced in Section 4.5. Equation (4.3) is the modified form of Eq. (3.1) that includes both spatial and time weighting of the learning data points used to estimate values of *Y* for the test data points. In this section we consider the effects of weighting on VR (Variance Reduction) using the data set generated with the program included in Figure 5.1. Since the artificial model used to generate the data is not time dependent, it makes no sense to time weight the data.

Therefore, the value of α in Eq. (4.3) was set to zero. The results are thus applicable to spatial weighting exclusively.

In Eq. (3.1) the *smoothing parameter* k is difficult to interpret. In the FKR program, an alternative definition of the smoothing parameter is used. To avoid confusion, the notation is changed to K. The definition of K is that the spatial weight of the furthest point (out of the *numnn* points used in the calculation) is $1/K$. We can express this as an equation:

$$1/K = e^{-(k*dsmax)} \qquad (5.6)$$

In this equation *dsmax* is the maximum dimensionless squared distance from the test point to the *numnn* learning points. The smoothing parameter k can therefore be determined from K:

$$k = ln(K)/dsmax \qquad (5.7)$$

The weights of all the *numnn* learning points are thus:

$$w(x_i, x_j, K) \geq 1/K \qquad (5.8)$$

In Figure 5.9 weighting w is shown for several values of K as functions of $D(i,j)$ (the normalized distance between the ith learning point and the jth test point). It should be clear that $D(i,j)$ is never used. Only the squared distance $D^2(i,j)$ is computed and used to determine the weights.

The interesting question is: how does the use of data weighting affect the results? Tables 5.11 through 5.13 were generated

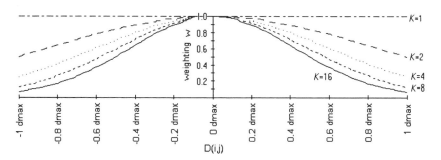

Figure 5.9   Weighting as a function of K.

TABLE 5.11 VR as a Function of h and K: fast mode, order = 0, Pure Signal

h	ppc	$K=1$	$K=4$	$K=16$	$K=64$
5	312.50	53.38	56.37	59.17	61.83
6	156.25	53.41	58.95	63.45	67.70
7	78.13	61.18	68.54	73.92	77.83
8	39.06	71.31	79.24	84.36	87.66
9	19.53	85.57	89.76	92.45	94.13
10	9.77	90.86	93.67	95.34	96.26

TABLE 5.12 VR as a Function of h and K: fast mode, order = 1, Pure Signal

h	ppc	$K=1$	$K=4$	$K=16$	$K=64$
5	312.50	76.76	80.47	83.26	84.43
6	156.25	89.30	91.81	93.54	94.75
7	78.13	92.76	94.85	96.18	97.04
8	39.06	96.04	97.23	97.95	98.39
9	19.53	98.21	98.68	98.97	99.14
10	9.77	98.83	99.02	99.14	99.21

TABLE 5.13 VR as a Function of h and K: fast mode, order = 2, Pure Signal

h	ppc	$K=1$	$K=4$	$K=16$	$K=64$
5	312.50	86.33	94.74	96.91	97.97
6	156.25	97.27	98.76	99.19	99.41
7	78.13	98.31	99.18	99.45	99.58
8	39.06	98.87	99.29	99.44	99.55
9	19.53	99.43	99.46	99.45	99.43

using the *fast mode* option, *nlrn*=10000, *ntst*=5000, and the correct model space (i.e., $X2$, $X5$, $X9$). These three tables are for the pure signal data column, and the results are predictable: VR increases as h increases, as the *order* increases, and as K increases. The effect of increasing K is to reduce the influence of distant points. For data uncorrupted by noise, this improves accuracy.

As the data becomes "noisier," the behavior is more complex. As h increases, the number of points per cell *(ppc)* becomes smaller, and eventually VR begins to decrease with increasing h. Tables 5.14 through 5.16 are similar to Tables 5.11 through 5.13 but for data that is 50 percent signal and 50 percent noise. Note in these tables that the optimum values of h tend to decrease as *order* increases. For example, for $K=1$, the best results obtained for *order* = 0, 1, and 2 are for h = 9, 7, and 6. What is implied here is that as *order* increases, more data

TABLE 5.14 VR as a Function of h and K: fast mode, order = 0, 50% Signal

h	ppc	$K=1$	$K=4$	$K=16$	$K=64$
5	312.50	26.31	27.86	29.34	**30.77**
6	156.25	26.02	29.01	31.43	**33.37**
7	78.13	29.58	33.51	36.28	**38.43**
8	39.06	34.00	38.08	40.58	**42.04**
9	19.53	41.31	43.02	**43.63**	43.54
10	9.77	40.51	**41.05**	40.05	38.20

TABLE 5.15 VR as a Function of h and K: fast mode, order = 1, 50% Signal

h	ppc	$K=1$	$K=4$	$K=16$	$K=64$
5	312.50	38.61	40.59	42.01	**43.06**
6	156.25	44.51	45.82	46.58	**47.00**
7	78.13	44.88	45.96	**46.34**	46.32
8	39.06	43.48	**44.00**	43.71	42.98
9	19.53	**38.71**	37.76	35.40	32.18
10	9.77	**11.88**	9.27	1.94	−9.54

TABLE 5.16 VR as a Function of h and K: fast mode, order = 2, 50% Signal

h	ppc	$K=1$	$K=4$	$K=16$	$K=64$
5	312.50	42.21	**43.86**	39.10	28.90
6	156.25	**46.74**	41.40	35.29	29.58
7	78.13	**42.78**	32.05	7.40	−29.5
8	39.06	**29.52**	8.90	−51.8	−142.5
9	19.53	**−54.49**	−142.7	−309.4	−467.5

points are required to get good estimates of Y. The noisier the data, the greater the need for data points for estimating Y.

The results in Tables 5.14 through 5.16 show the best VR for each value of h in boldface type. Note that as h and *order* increase, the optimum value of K decreases. The smaller the value of *ppc*, the less the incentive for reducing the effect of distant points. One might ask the question: if we make a better choice of data points, can we achieve higher VR? The answer is yes, and evidence justifying this claim is included in Table 5.17. The greater the value of *numcells*, the greater the VR. However, as shown in Section 5.5, the use of *numcells* causes a sharp increase in compute time. Another important point to note regarding *numcells* is that the best results are obtained using equal weighting (i.e., $K=1$). When a search is made for the nearest neighbors, these results indicate that reducing the weight for distant points is counterproductive.

Extending the analysis to much noisier data, results are included in Tables 5.18 to 5.20 for data that is 10 percent signal and 90 percent noise. Once again we note that the best results are achieved for *order* = 2 (VR = 8.06 for h = 3). One might get the mistaken impression that we should always use *order* = 2 because the best results are obtained using this algorithm. In Section 5.7 the choice of algorithm is considered, and it will be seen that the choice is affected by *nlrn* (the number of learning points). In this section the value of *nlrn* was held constant at 10,000; this value is large enough to yield the best results for *order* = 2 for the three-dimensional signal in the test data.

The weighting scheme suggested by Eq. (5.6) uses the same smoothing parameter for each dimension in a space. There is no

TABLE 5.17 VR as a Function of *numcells* and *K*: h = 8, *order* = 2, *numnn* = 39, 50% Signal and 50% Noise

numcells	K=1	K=4	K=16	K=64
1	29.52	8.90	−51.8	−192.5
2	31.00	14.03	−28.2	−93.4
5	34.64	25.48	8.85	−14.7
10	39.04	33.56	25.31	10.60
all	39.30	35.49	27.66	15.47

TABLE 5.18 VR as a Function of h and K: fast mode, order = 0, 10% Signal

h	ppc	$K=1$	$K=4$	$K=16$	$K=64$
5	312.50	4.97	5.30	5.61	**5.91**
6	156.25	4.51	5.22	5.76	**6.14**
7	78.13	4.67	5.47	5.98	**6.25**
8	39.06	4.10	4.88	**5.08**	4.93
9	19.53	**4.61**	4.37	3.30	1.72
10	9.77	**−0.45**	−1.80	−5.17	−9.50

TABLE 5.19 VR as a Function of h and K: fast mode, order = 1, 10% Signal

h	ppc	$K=1$	$K=4$	$K=16$	$K=64$
4	625.00	6.13	6.68	7.11	**7.42**
5	312.50	7.09	7.54	7.76	**7.85**
6	156.25	7.19	**7.49**	7.47	7.26
7	78.13	5.04	**5.20**	4.78	4.02
8	39.06	**0.07**	−0.10	−1.34	−3.14
9	19.53	**−10.5**	−12.7	−17.4	−23.5
10	9.77	**−61.8**	−66.0	−79.5	−100.6

TABLE 5.20 VR as a Function of h and K: fast mode, order = 2, 10% Signal

h	ppc	$K=1$	$K=4$	$K=16$	$K=64$
3	1250.00	7.34	**8.06**	7.42	6.38
4	625.00	**6.77**	6.12	5.03	2.58
5	312.50	**5.76**	1.64	−9.3	−28.6
6	156.25	**4.04**	−5.26	−18.7	−29.5
7	78.13	**4.12**	−24.4	−69.4	−136.0
8	39.06	**−29.3**	−67.2	−177.9	−437.3
9	19.53	**−183.8**	−346.0	−650.5	−939.9

theoretical reason why this scheme cannot be generalized. A more general form of Eq. (3.1) can be used:

$$w(x_i, x_j, k) = e^{-f(k)} \qquad (5.9)$$

In this equation k is now a vector and $f(k)$ is defined as follows:

$$f(k) = \sum_{d=1}^{d=p} k_d D_d^2 \qquad\qquad (5.10)$$

In this scheme, each dimension in the p-dimensional space is given a different smoothing parameter (i.e., k_d, d = 1 to p). Each value of k_d is multiplied by $D^2{}_d$, which is the dimensionless squared distance in the d-direction between points i and j. The problem with adopting Eq. (5.10) is that it vastly increases the number of free parameters and therefore degrees of freedom in the search for a model. The computational time increases linearly with the number of combinations of the k vector that are considered. For example, assume we choose only three values of k_d in each direction: the total number of combinations would be 3^p. For a 3D space we would thus have to consider 27 combinations, and for a 4D space the number of combinations would be 81. Thus for 4D analyses the computational time would be increased by almost two orders of magnitude! For the particular data sets considered in Tables 5.11 through 5.14, we see that the effect of weighting is significant but not crucial. It is doubtful that searching each space using different smoothing parameters in each direction would radically alter the results.

5.7 COMPARING THE THREE ALGORITHMS

A crucial distinction must be made when considering the choice of algorithm: Are we concerned with modeling, or are we actually making predictions? The modeling task requires considering many spaces and comparing predicted values of y with actual values of y for many test points. This is a compute-intensive activity, and computational speed is an important consideration. However, once a model space has been located and all that is required is a single prediction, then speed is no longer an issue. The choice of parameters (including the *order* of the algorithm) is made entirely on the basis of accuracy.

When the analyst's concern is to optimize the software for making predictions, the choice of the main parameters (i.e., *order* of the algorithm, treeheight *h*, number of nearest neighbors *numnn,* number of adjacent cells *numcells,* and weighting parameters K and α) is quite simple. All that is required is some experimentation varying the parameters in a final pass using the best model (or models). For each model the combination of parameters yielding the best result is then selected and used for making future predictions. For greatest accuracy the value of *numcells* is invariably set to a large value so that all adjacent cells are examined in the search for the *numnn* nearest neighbors.

The choice of parameters for modeling is more complicated. This selection process is a matter of compromise: the choice must yield meaningful results and at the same time must be computationally efficient. As the size of the data set and the number of candidate predictors *ncp* increase, the choice of parameters becomes more crucial. For large problems the *fast mode* option is usually selected. Usage of this option simplifies the selection of modeling parameters: *numcells* = 1, *numnn* is just the number of points in each test cell, and K is set to 1 (i.e., all points in the cell are equally weighted). The selection process thus narrows down to the choice of *order* and h.

In the previous sections the following picture emerges: the best accuracy can be obtained using the highest order algorithm (i.e., *order* = 2) but only if enough points are available for determining the constants needed to define the surface. As the noise component in the data increases, the number of data points required to obtain acceptable accuracy increases. For problems with little noise, the best results are observed for *order* = 2 and fairly large values of h. As the noise level increases, the optimum value of h is generally smaller. As *nlrn* (number of learning points) decreases, the optimum value of h tends to decrease, and the best results are observed for *Order 0* or *1* rather than *Order 2.*

To illustrate this point, the values of *nlrn* (number of learning points) and h were varied using each *order* algorithm. Using the data set which is 10 percent signal and 90 percent noise, the values of h and VR for which the best results were obtained are summarized in Table 5.21. The value of *ntst* (number of test points) was 5000 for all the experiments. For *nlrn* >= 3000, the best results are noted for *Order 2*. For *nlrn* <= 2000 *Order 0*

TABLE 5.21 Values of h Yielding Best VR as Functions of $nlrn$ and $order$ (0, 1, and 2).

$nlrn$	$h(0)$	$VR(0)$	$h(1)$	$VR(1)$	$h(2)$	$VR(2)$
10000	4	5.37	6	7.11	3	**7.34**
8000	4	5.26	5	6.74	3	**6.99**
6000	3	5.30	5	5.90	3	**6.78**
4000	3	5.39	5	4.90	3	**5.79**
3000	3	5.12	4	4.47	3	**5.51**
2000	3	**5.29**	4	4.19	3	4.50
1500	3	**4.84**	3	3.39	3	3.21
1000	3	**4.36**	3	2.12	3	0.40

Boldface shows best results in row.

yields the best results. For *Order 0* and *1* the optimum values of h decrease as $nlrn$ decreases. The fact that the best results for *Order 2* are all observed at $h = 3$ is a result of the particular signal function included in the program used to generate the data (i.e., Figure 5.1). In the three-dimensional signal space (i.e., $X2$, $X5$, $X9$), there are four maxima and four minima. This sort of function is easily modeled after being separated into eight separate regions (i.e., $h = 3$) and then using complete second-order multinomials as the fitting functions.

These results are valid only for the particular data set used to obtain the results. They do, however, indicate qualitatively the sort of behavior one can expect. If, for example, the true signal was four dimensional rather than three dimensional, the point at which the *Order 2* values of VR are less than *Order 0* and *1* would be expected to occur at a higher value of $nlrn$. The higher the dimensionality d of the model, the greater the number of coefficients required to describe the surface in each cell (i.e., $1 + d + d*(d + 1)/2$ for *Order 2* modeling), and therefore the greater the number of learning points needed to achieve comparable accuracy. In summary, we conclude that the greater the value of $nlrn$, the more likely it is that we can use the *Order 2* Algorithm for modeling. However, if $nlrn$ is small, then more than likely the best results will be obtained using *Order 0*. The use of *Order 1* is best at some intermediate range. However, some experimentation is required to determine the best choice of *order* and h.

NOTES

1. TIMES (Insightware Ltd, Haifa, Israel, *www.insightware.com*).

2. FKR (Fast Kernel Regression) program (Insightware Ltd., Haifa, Israel, *www.insightware.com*).

3. B.W. Kernighan and C.J. Van Wyk, *Timing Trials: Experiments with Scripting and User-Interface Languages, http//kx.com/a/kl/examples/bell.k,* 1997.

6

MODELING STRATEGIES

6.1 THE MODELING PLAN

Any serious modeling effort requires careful planning. The objectives of the project must be specified and then some thought must be given to ways of achieving the objectives. Two broad classes of modeling projects are *prediction modeling* and *filtering*. Some aspects regarding the planning processes for these two types of projects are similar, and yet there are differences.

The term *prediction modeling* is self-explanatory: it is used to describe projects in which the aim of the modeling process is to provide models for making predictions. Many options are available, and the analyst must make a number of decisions in the planning stage:

1. The variable (or variables) to be predicted.
2. The measures by which the models will be evaluated.
3. The database that will be used for the process.
4. The candidate predictors.
5. The modeling technique (or techniques) to be used.
6. Partitioning the available data records.

The first step is to identify exactly what we wish to predict. It's not enough to say that we want an S&P model. What sort of time horizon are we talking about? Are we trying to predict the change from today's close to tomorrow's close, or are we interested in an intraday model? Perhaps we are interested in longer-term predictions, or perhaps we are only interested in predicting volatility. Many options are available, and it is important to identify the objective at an early stage in the planning process.

Many measures of performance can be used to evaluate a model. In the previous chapter, VR (Variance Reduction) was used as the primary measure. However, it should be clear that the analyst might prefer other measures of performance. In Section 6.6 several alternatives are discussed.

An important step in any modeling project is to gather and check the data. This is often a nontrivial task, and it is linked with selecting the candidate predictors. The choice of candidate predictors is typically based on the available data. However, if the selection of candidate predictors precedes construction of the database, the analyst must then acquire the necessary data.

The choice of the modeling technique (or techniques) is usually based on two factors: familiarity and availability. Most analysts are familiar with at least one method and thus tend to use their favorite method. Availability of user-friendly software is also crucial. Thus, the introduction of a new modeling technique must take these factors into consideration. Compelling arguments are required to convince analysts to try something new, and user-friendly software based on the new method must be available.

Use of the data records should be considered prior to initiating actual computer runs. Because most methodologies use the learning-test paradigm, the first question is, How should one partition the data? If there is sufficient data, it is always worthwhile to leave some data aside for final evaluation. This subject is discussed in Section 6.2.

The term *filtering* is used to describe a modeling activity designed to improve models by filtering out probable "losers." Filtering is usually applied to trading systems in which a reasonable number of trades (either actual or simulated) are available. One approach to filtering is to list all the trades with the Y variable as the trade outcome. The modeling process is then directed toward predicting Y as a function of a subspace of the candidate predictors. If a subspace can be located in which there are regions with high densities of winners and other regions with high densities of losers, then all one has to do is to wait for a signal from the trading system. At that point one should determine the region in which the relevant predictors fall and then act accordingly. This concept is illustrated in Figure 6.1.

This figure shows that the results of the trades are partitioned into four regions in the 2D subspace created with predictors $X17$ and $X35$. The region marked A has mostly winning trades, the D region has mostly losing trades, and the B and C regions have a comparable number of winners and losers. For this particular example, the number of trades is small so it would be difficult to come to any statistically significant conclusion. However, let's suppose that approximately the same ratio of winners to losers per region persists through hundreds of trades. We would then have an invaluable filter. Consider, for example, a situation in which the trading signal generator issues a signal, and the values of $X17$ and $X35$ are at that moment in region A (i.e., the value of $X35$ is less than 1.7 and the value of $X17$ is greater than -4.3). For this situation, we would have a high expectation of a winning trade. Alternatively, if the values of $X17$ and $X35$ put us in region D, we will clearly reject the trade. Whether or not a trade is accepted or rejected when the values of $X17$ and $X35$ put us in regions B or C is not

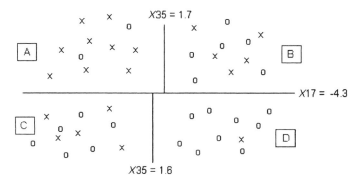

Figure 6.1 An example of a filtering application. The x's represent winners and the o's represent losers.

clear from this figure. This decision would be based on a more detailed analysis regarding trades in these regions.

The major difference in *prediction modeling* and *filtering* is in the choice of measures of performance. Typically, the selection process for filtering projects is based on analysis of the trading results for the trading system with and without filtering. For prediction modeling, more straightforward measures such as VR are usually sufficient. The choice of candidate predictors might or might not be different, and this is the most difficult task in any modeling project. The success or failure of any type of modeling project is based primarily on the choice of candidate predictors. Within the chosen set of candidate predictors, do any combination (or combinations) contain real information that will shed some light on future performance?

6.2 OUT-OF-SAMPLE TESTING

When a modeling effort yields a "useful" result (e.g., VR), the analyst should ask the following questions:

1. Are there any fundamental flaws in the methodology used to obtain the model?
2. Are the results just a statistical artifact?
3. If the model is real, how *persistent* is it? In other words, can we expect it to hold up for a reasonable amount of time into the future?

The function of *out-of-sample* testing is primarily to provide an answer to the third question. If there are sufficient data records and if the *Y* column (the data column being modeled) is just random noise, the probability that the models (or model) will show encouraging results is low. This point is discussed in greater detail below.

The concept of out-of-sample testing is to leave some data out of the modeling process and then use it for testing in the final phase of the project. If the results show that the performance continues to be acceptable, then one can proceed towards actual usage of models. If the results are poor, one can only conclude that either the modeling methodology was flawed or models that worked during the modeling period were no longer valid during the out-of-sample (or evaluation) period. A very rudimentary flowchart of the process is shown in Figure 6.2. Consider the parameter *num_best_models* as a user-supplied parameter. One does not necessarily have to use all *num_best_models*. However, supplying details regarding the best models is a useful software feature that allows the user greater scope in deciding how to use the results of the modeling process.

The need for out-of-sample testing is greater for noisy problems. When the signal-to-noise ratio is low, the difference between a successful model and a useless model is small. A few percentage points change in the VR (Variance Reduction) of the model can be the difference between a model with predictive power and one with negligible value. For example, assume that we have a data set that is purely random. What is the probability of finding some combination of parameters that exhibit VR of say 5 percent? In Appendix B an equation is derived for the

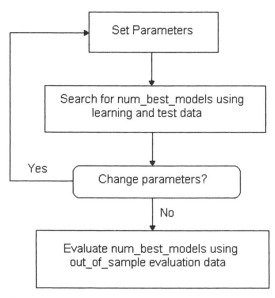

Figure 6.2 Basic flowchart of the modeling process.

distribution of VR that can be expected under certain conditions (i.e., random data, *nlrn* = *ntst, order* = 0 and *fast mode* of operation). For these conditions, the expected value of VR/100 (i.e., $E(VR/100)$) is as follows:

$$E(VR/100) = -\frac{N_t}{N_t - 1}(\mu_t - \mu_l)^2 - \frac{N_c - 1}{N_t - 1}\left(\frac{\sigma_l}{\sigma_t}\right)^2 \qquad (6.1)$$

In this equation, N_t is the number of test points (i.e., *ntst*) and also *nlrn*; μ_t and μ_l are the test and learning mean values; and the σ's are the standard deviations. The constant N_c is the number of cells and is related to the treeheight (i.e., $N_c = 2^h$). As the number of test and learning points become large, the assumptions that $\mu_t = \mu_l$ and $\sigma_t = \sigma_l$ are reasonable, so Eq. (6.1) can be written in its asymptotic form:

$$E(VR/100) \approx -\frac{N_c - 1}{N_t - 1} \qquad (6.2)$$

For example, assume we have 2000 data points equally divided between test and learning and use a tree of h = 4 (i.e., 16 cells). If there is no information in the data, we would expect values of VR = 100($-15/999$) = − 1.5%. The expected value of the variance of VR is also derived in Appendix B:

$$\sigma^2_{VR/100} \approx 6(N_c - 1)/(N_t - 1)^2 \qquad (6.3)$$

For the example, the value of σ_{VR} = 100 * *sqrt*(6*15)/999 = 0.95%. The two sigma limits for VR are thus: −3.4 <= VR <= 0.4. In other words, the probability of VR reaching 5 percent is very small. (The distance between the expected value of −1.5 percent and 5 percent is 6.84σ.)

Equations (6.2) and (6.3) are applicable only to *Order 0* and are not functions of the dimensionality d of the space. However, for the higher order algorithms, d is important. Table 6.1 includes results from the following experiment. Eleven columns of random data were generated. The first 10 columns were con-

TABLE 6.1 Average VR (and σ_{VR}) for Combinations of *Order* and Dimensionality

h	spaces	order 0	order1	order2
4	all spaces (175)	−0.33 (±0.24)	−1.62 (±0.63)	−4.61 (±1.80)
4	1D spaces (10)	−0.30 (±0.30)	−0.76 (±0.33)	−1.19 (±0.32)
4	2D spaces (45)	−0.41 (±0.21)	−1.38 (±0.54)	−2.92 (±0.83)
4	3D spaces (120)	−0.29 (±0.23)	−1.79 (±0.60)	−5.54 (±1.24)
5	all spaces (175)	−0.83 (±0.36)	−3.40 (±1.10)	−10.46 (±3.88)
5	1D spaces (10)	−0.65 (±0.24)	−1.60 (±0.42)	−2.69 (±0.70)
5	2D spaces (45)	−0.87 (±0.42)	−2.71 (±0.91)	−6.35 (±1.20)
5	3D spaces (120)	−0.83 (±0.34)	−3.80 (±0.91)	−12.65 (±2.27)

sidered as candidate predictors, and the eleventh column was assumed to be the *Y* column. All combinations of one, two, and three dimensions were tested using all 10 candidate predictors with 3200 learning and 3200 test points. The treeheight was first set to 4 and then to 5, so that there were 16 (and then 32) cells and 200 (and then 100) learning points per cell. For each *order* the number of spaces examined was 175 (i.e., 10 1D spaces, 45 2D spaces, and 120 3D spaces.) The average values (and standard deviations) of VR for the spaces are summarized in Table 6.1.

Since many more 3D combinations were tested, the averages for all spaces are biased toward the 3D averages. However, the results show the effect of *d* and *order* on VR when the data is void of information. Increasing *order* makes the expected value of VR increasingly negative, and dimensionality has an increasingly important effect as *order* increases. As one would expect based on Eq. (6.2), the $E(VR)$ becomes increasingly negative as *ppc* (the number of learning points per cell) decreases (i.e., *h* increases). Note that for a fixed value of *nlrn,* as *h* increases, $N_c = 2^h$ increases and $ppc = nlrn/N_c$ decreases. An additional experiment was run in which *h* was increased to 6 and *nlrn* and *ntst* were increased to 6400. The value of *ppc* for this experiment is thus the same as for *h* = 5 in Table 6.1. The results were comparable for all three values of *order,* thus confirming that *ppc* is the dominant parameter.

The results in the table indicate that the probability of obtaining useful Variance Reduction when the data is purely random is very small. So one might ask the question: why bother with out-of-sample testing? The more probable danger is that a model exists at one period in time but due to the dynamic nature of the markets, it eventually breaks down. Out-of-sample testing is one method for measuring the *persistence* of a model. Strategies for modeling dynamic systems (such as financial markets) are discussed in the next section.

Some problems are fundamental and usually are not caught in out-of-sample testing. One such problem is concerned with misalignment of data. If, for example, we are trying to develop a model for predicting the one-day change in some financial instrument, then we must make sure that the ith record contains only candidate predictors that are really known at the time i. Similarly, the Y column (which for this example is the one-day price change) must be created as the price at time $i + 1$ minus the price at i. If the indexing is off by one and the difference included for the ith record is the price at i minus the price at $i - 1$, then the results are usually wildly optimistic. When the initial results seem unreasonably high, this potential error should be the first possibility to be investigated.

Another more subtle error is a partial overlapping of the Y values. For example, assume we are trying to develop a model for predicting the three-day change in some financial instrument. A database of daily records is created in which there are X's (the candidate predictors) and the Y column (the three-day changes). If all the previous records are used to predict Y for the ith record, then the Y values for records $i - 1$ and $i - 2$ contain information not known at time i. What is required is a *gap* in the modeling scheme such that only records for which the Y values are known at time i are used to predict Y at time i.

6.3 MODELING DYNAMIC SYSTEMS

In addition to all the complexities associated with financial market modeling, the dynamic nature of the markets must also be considered. For such systems, we seek models that evolve over time. This feature is one reason why neural networks have become so popular. Neural networks have the ability to "learn." Typically, learning is accomplished by periodically updating the weights used within the network as new data points become available. Theoretically, a particular input variable that is important at one stage in time can gradually lose its importance as new data is obtained. Conversely, an unimportant input variable can gain in importance. The gain or loss in importance of an input variable is measured by the weights associated with the variable within the network.

To add some degree of a dynamic nature to a KR model, several software features have been suggested:

1. Use of *time weighting* of the data (see Section 4.5).

2. Use of either the *growing* or *moving window* options (see Section 3.5).

The problem with both of these features is that they change the data (or the data weights) used to make predictions, but they don't really change the inputs to the models. Once a KR model has been selected, its inputs are fixed. What we require for dynamic systems such as financial markets is some way of periodically replacing models.

To add a dynamic flavor to KR modeling, one approach is comparable to the *moving window* option but is based on all the data (i.e., learning, test, and evaluation data sets). The values of *nlrn, ntst,* and *nevl* are first chosen. Using the learning and test data sets (i.e., the first *nlrn* plus *ntst* data records), the *num_best_ models* are selected. These models are then tested using the next out-of-sample *nevl* evaluation data records. Let us call this phase of the analysis the first *fold.* To proceed to the next fold, all data sets are incremented by *nevl* records, and the process is repeated *num_folds* times. Note that within each fold it is possible to use either the *static, growing,* or *moving window option.* A schematic view of this process is shown in Figure 6.3.

Note in this figure the introduction of several new parameters: *startlrn, starttst,* and *startevl.* These are pointers to the starting records of the learning, test, and evaluation data sets.

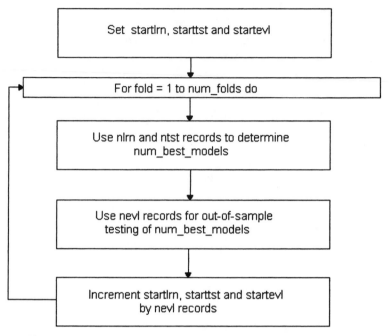

Figure 6.3 Basic flowchart for modeling dynamic systems.

Typically, the three data sets are consecutive; therefore the following relationships are used to increment these parameters:

$$starlrn = startlrn + nevl \qquad (6.4)$$

$$starttst = startlrn + nlrn \qquad (6.5)$$

$$starevl = startltst + ntst \qquad (6.6)$$

The scheme shown in Figure 6.3 can be generalized if the values of *nlrn, ntst,* and *nevl* are allowed to vary from fold to fold. This generality is required if the system being modeled includes multiple data records per unit of time. For example, assume that a stock selection system is being modeled. Further, assume that the database includes monthly price changes in the stocks and that the number of stocks being followed varies with time. Stocks are constantly being added and dropped from the database. For this particular modeling project, it is more natural to specify *nlrndates* (the number of dates in which all records are considered as belonging to the learning data set), *ntstdates,* and *nevldates.* Once the number of dates constituting each data set is specified, the values of *nlrn, ntst,* and *nevl* can be computed for each fold. Equations (6.4) to (6.6) must also be modified. The *startlrn* pointer is increments based on *nevldates.* Similarly, the *starttst* and *startevl* pointers are incremented based on *nlrndates* and *ntstdates.* The modified flowchart is shown in Figure 6.4.

What can be expected from the schemes shown in Figures 6.3 and 6.4? Most probably, if the total modeling period is large, and if the number of candidate predictors is much larger than the maximum dimensionality specified for the modeling process, the *num_best_models* will vary from fold to fold. In particular, the predictors appearing in the best models will change from fold to fold. For example, let's say that the best model in

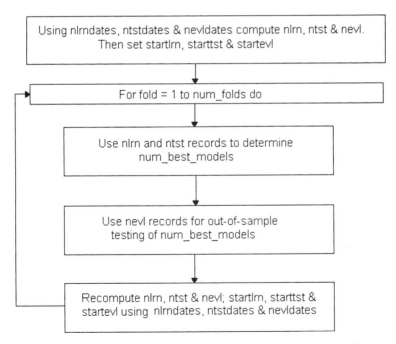

Figure 6.4 Flowchart for dynamic systems with varying number of records per unit of time.

fold 1 is found to be the space created from predictors $X9$, $X46$, and $X85$. These predictors might not appear in any of the best models in fold 2. Sometimes, however, one predictor is particularly powerful, and it might appear in several of the *num_best_models* and even in subsequent folds. What one desires is that useful models (i.e., models that yield sufficient out-of-sample MC (modeling criterion) such that they can be usefully exploited) are found in all or at least many of the folds.

This concept was used in an actual dynamic stock selection project. The total number of stocks included in the database was over 7000, and the analysis was based on monthly results from April 1990 to July 1998. The quantity being modeled was the excess return for a stock. For example, assume record i is created from data for stock j in month k. The excess return is the difference in the value of the stock from month k to month $k + 1$ minus the average difference for all the stocks in the same time period. If this value is positive, then stock j was better than average during this time period.

Table 6.2 includes values of *nlrn, ntst,* and *nevl* in each of seven folds of the data. Note that the values increase from fold to fold because over time more stocks were being added to the database than were being removed. The values of *nlrndates, ntstdates,* and *nevldates* were each 12. In other words, one year of data was used as learning data and the test data was the next year of data. (Note that for the final fold, even though *nevldates* = 12, the available number of remaining dates was only 4.) Once the *best_models* were determined for each fold, the records from the learning and test data sets were combined to

TABLE 6.2 Values of *nlrn, ntst,* and *nevl* for Stock Selection Analysis

Fold	Startlrn	Starttst	Startevl	Nlrn	Ntst	Nevl
1	4/90	4/91	4/92	16832	18104	19719
2	4/91	4/92	4/93	18104	19719	20378
3	4/92	4/93	4/94	19719	20378	23454
4	4/93	4/94	4/95	20378	23454	25785
5	4/94	4/95	4/96	23454	25785	28656
6	4/95	4/95	4/97	25785	28656	29762
7	4/96	4/97	4/98	28656	29762	10219

make predictions on the data in the third year. For subsequent folds, the process was advanced by one year.

For this particular analysis, the model is updated yearly. However, there is no reason why the updating can't be performed more frequently. The limiting value on the updating frequency is the "acceptable" number of data records used in the evaluation data set. For the analysis summarized in Table 6.2, we see that the number of evaluation data records in each fold is quite large and therefore the frequency can easily be increased. Even if we produce a new model every month, the number of evaluation records will be in the 1500 to 2500 range, and this should yield reasonable statistics.

6.4 CROSS-SECTIONAL MODELING

When one considers time series, the typical data structure is an array of records in which the date (or date/time) of record i is less than that for record $i + 1$ and greater than that for record $i - 1$. There are many problems in financial market modeling for which this data structure is insufficient. A more general data structure allows multiple records for a given time slot. Furthermore, if the number of records per time slot is variable, then we have a data structure that is useful for most types of modeling problems encountered in the financial community. Projects that require multiple records per time slot fall into the category of *cross-sectional modeling*.

An example of a cross-sectional analysis was discussed in Section 6.3. The application was based on a stock selection project in which the variable being modeled was the one-month excess return. Over 7000 stocks were included in the database, and the data consisted of monthly results for 100 months (April 1990 to July 1998). The number of stocks included in the database varied from month to month because stocks were being added and removed from the database monthly. The analysis used the *growing* option described in Section 3.5, with a modification required to accommodate the cross-sectional data structure.

If the *growing* or *moving window* option is used, the simple scheme described in Section 3.5 is inadequate. In Section 3.5, immediately after a test record has been used, it is then added to the learning tree. For cross-sectional modeling, a record can be used only when it can be considered as belonging to the past. Thus for the stock selection project described above, records from a particular month can only be added to the learning tree after all stocks from that month have been tested. At that point, all the stocks from that particular month are added to the learning tree, and the test then proceeds to the next month. Similarly, if the *moving window* option is used, the block of stocks from the earliest remaining month in the tree are removed at that same point before proceeding to the next month.

Some modeling projects as originally defined lack a sufficient number of data records. Cross-sectional modeling is sometimes an attractive alternative that can be used to greatly increase the number of records. Typically, projects using daily data are faced with this problem. For example, assume we are trying to create a model for predicting the daily price change in a particular futures market. Assuming about 250 trading days per year and assuming that we have 10 years of available data, the number of available records is about 2500. If we divide the data into learning, test, and evaluation sets, the number of records per set is small. Furthermore, considering the dynamic nature of the futures market, we would like to use the *folds* concept discussed in Section 6.3 and that would further reduce the sizes of the data sets. If we don't use the *folds* concept, we are forced to use data that stretches as far back as 10 years for making predictions. However, if we do use the *folds* concept, we have very little data per fold.

One possible solution to this problem is to find a single model that yields decent results for a group of similar financial instruments. For example, instead of trying to model a single futures market, one might attempt to model a group of futures markets. Such a change requires a degree of coordination. The individual candidate predictors must, of course, be normalized. For example, fractional changes in prices rather than actual changes in prices must be used. Slopes and moving averages must also be based on relative rather than absolute quantities. Consider the advantages of such a scheme. Assume that we include data from

20 different futures markets. This would increase our database to about 50,000 records. Having about 5000 records per year would allow a scheme in which one year of data could be used as the learning data set, followed by one year of test data and then one year of out-of-sample evaluation data. This scheme would result in eight folds of data and should yield results that would be a real test of the efficacy of the technique. If, for example, the out-of-sample results for most of the folds were "good," one would have some faith in the methodology and there would be justification for incorporating the resulting model into a trading system. At the end of every year, the model would be updated using only the past three years of data.

6.5 COMBINING MODELS

In the previous sections, the emphasis was on finding *num_best_models* rather than a single best model. If, for example, we can determine several different models that yield a few percent VR, the question arises: Can we combine them in such a way as to determine a single better model? (Let's call the new model a *super-model.*) If the models are highly correlated, then we cannot expect significant improvement. For example, assume each of the models yields very good predictions for one particular period within the total data range and is useless for the remainder of the time period. If all the models yield good predictions during the same period, then nothing will be gained by combining them. However, if the "good" periods don't overlap, then we can expect an improvement by combining the models.

The technique for combining models is straightforward:

1. Create a new set of candidate predictors that is the outputs from the models to be combined.

2. Use any modeling technique to create a new model based on these new candidate predictors.

There are many modeling techniques available for creating a new model. The simplest method is to just take the average values of the predictions from the best models and use these average values as the predictions of the combined model. This approach to modeling is discussed by Armstrong in his book on Long Range Forecasting.[1] He refers back to earlier work by Bates and Granger on Combining Forecasts.[2] In the stock selection case study included in Section 7.6, a significant improvement is noted by using the average of the 5 best models as compared to just the single best model.

More sophisticated methods for combining models can also be used. For example, one might consider using kernel regression to combine the outputs of the best models into a single super-model. An alternative is to use a neural net approach to modeling. The neural network approach to modeling is particularly effective when the number of input parameters is not large.

A flowchart of the process is shown in Figure 6.5. Although the flowchart shows the entire process in a single loop, it is more likely that the modeling process will be broken down into two loops. In the first loop, *num_best_models* are determined. When the outputs from these models become available, then the super-models (i.e., the super-model in each fold) can be generated in a separate loop. Note that generation of a super-model can be based on any modeling technique. The split between learning, test, and evaluation data sets can be the same in both blocks. If the results from the out-of-sample testing are successful, one can either continue testing using new data as it

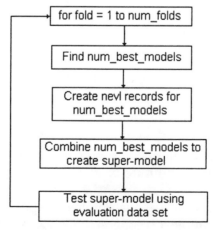

Figure 6.5 Creating a super-model from num_best_models.

becomes available or actually use the model in a real trading environment.

6.6 MEASURES OF PERFORMANCE

Several *measures of performance* were discussed in Section 1.4. When initiating a modeling project, the choice of the measure (or criterion) to be used is usually considered in the project planning stage. If *ncp* (the number of candidate predictors) is large, one usually chooses a measure that can be computed quickly. However, the project can be designed to include several modeling stages. The first stage might be directed toward reducing *ncp* to a more manageable number using a quick measure of performance such as VR (Variance Reduction). In the second stage, a more costly measure of performance might be used with the remaining candidate predictors.

A useful rule-of-thumb is to choose a criterion that is reasonably close to the end-use of the models. For example, consider the stock selection application discussed in Section 6.3. The purpose of the project was to choose two groups of stocks each month: one group that could be expected to outperform the market over the next month and another that could be expected to be underperformers. The idea behind this project was to help develop a *market-neutral* portfolio of stocks. The use of VR for such a project doesn't really answer the project requirement. It is not necessary to be able to make predictions over the entire range of stocks. What is required is a model in which the group of stocks that exhibit the highest predictions over the next month outperform the group exhibiting the lowest predictions.

To determine the two groups, the N records are first sorted by increasing values of $Ycalc$ (the predicted values of Y). The number of records N is either the number of test records or the number of evaluation records, depending on which data set is under consideration. The number of records per group n is equal to $P*N/100$. The user-supplied constant P is the percentage of N records in each group. The following measure of the value of a model is made by first computing the averages and standard deviations of Y of the two groups of stocks. We can define SEP (separation) of the model as follows:

$$SEP = (\overline{Y}_1 - \overline{Y}_2) / \sqrt{\sigma_1^2 / n + \sigma_2^2 / n} \qquad (6.7)$$

The subscripts 1 and 2 refer to the two groups: group 1 is the group of best performers, and group 2 is the group of worst performers. The terms *best* and *worst* refer to the predicted values of Y. Some of the worst performers based on predicted values of Y might outperform some of the best performers. If the value of n is reasonably large, and if there is no relationship between the actual values of Y and the predicted values (i.e., *Ycalc)*, one can expect SEP to be normally distributed with mean 0 and standard deviation of 1.[3] The sign of this measure is definitely meaningful. Negative values of SEP imply that on average the worst groups of records (based on the values of *Ycalc*) outperform the best group. What one would like to see is a large positive value of SEP. If the normal distribution is valid, the probability of getting a value of SEP greater than 2.33 is 1 percent if the model is purely random. The probability of getting a value of SEP greater than 3.08 is 0.1 percent.

The problem with using SEP for initial scanning of the data is the significant increase in computer time when compared to using VR. The separation of the values of *Ycalc* into two groups requires a sort, and if N is large the sort time is $O(N\log(N))$. For the stock selection and similar problems, the choice of P can be made on the basis of how many stocks (or comparable items) one wants in each group. The significance of the value of SEP can be seen at a glance. However, significance and usefulness should not be confused. The remaining question is how does one convert the predictions into a successful trading system? One would have to propose rules as to how the trading system would operate and then test it using software that can simulate its usage in the proposed trading system. This subject is considered in Chapter 7.

Another statistic that is more meaningful when the group concept is applied is *FracSS* (Fraction Same Sign). If *FracSS* is measured only for the test or evaluation records falling within the two groups, the weakness of this statistic is overcome. The problem with *FracSS* when applied to all predictions is that values of *Ycalc* close to zero are treated the same as the extreme values. Since one tends not to use predictions close to zero, use of them in the computation of *FracSS* simply complicates interpretation. A test of significance for *FracSS* is included in Appendix E.

Another well-known statistic that requires sorting of the values of *Ycalc* is the Spearman Rank Correlation Coefficient.[3-6] Both SR (Spearman rank) and CC (the correlation coefficient defined in Eq. 1.16) are measures of the relationship between the values of *Y* and *Ycalc*. A value of 1 for CC implies that all values of *Y* and *Ycalc* fall on a straight line with a positive slope. A value of −1 implies that the slope is negative. A value of 1 for SR implies that the values of *Y* corresponding to the sorted values of *Ycalc* all trend upward but not necessarily on a straight line. A value of −1 implies that the trend is negative. For purposes of trading, it is more meaningful to measure the trend rather than the proximity to a straight line. A schematic representation of SR is shown in Figure 6.6.

An option that is always available to the analyst is to consider the end-usage of the model and choose a measure of performance based on this end-usage. If, for example, the model is to be incorporated into a trading system, and if the rules of how one would use the model can be specified ahead of time, then trading performance can be used as the basis for model selection. The major problem with this approach is that added degrees of freedom based on the trading rules become part of the modeling process. For every space considered, if many combinations of entry and exit rule parameters are tried, the probability of overfit becomes a major problem. By selecting the

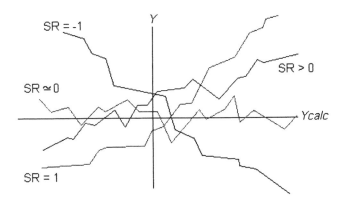

Figure 6.6   Several different curves for *Y* versus *Ycalc*. SR is Spearman Rank Correlation Coefficient.

parameters optimally, we can artificially obtain excellent results that catch a few big moves. The probability that these optimal parameters will actually perform well in the future is not great. Out-of-sample testing should reduce this danger, but nevertheless, the greater number of free parameters included in the modeling process, the greater the probability of overfitting the data. My personal preference is to separate the modeling process from the final task of creating a trading system.

If one decides to use a trading-based measure of performance, the next question to be answered is what trading-based statistic should be used. There are a number of possibilities, and the usual choice is a measure that maximizes a reward-to-risk ratio. Usually, the "reward" is the return or excess return of the simulated trading system. There are a number of well-known measures of "risk." The popular *Sharpe Ratio* measures risk as the annualized daily fractional equity changes (Eq. 2.11). (In a provocative article, Benoit Mandelbrot argues that using the standard deviation of equity changes underestimates the real risk due to the chaotic nature of the financial markets.)[7] Another measure of risk used in the *Advocom Ratio* is the root-mean-square drawdown throughout the measurement period. Yet another useful statistic that also measures a reward-to-risk ratio is the *Profit Factor*. The definition of this ratio is the total amount earned in winning trades divided by the total amount lost in losing trades. Most trading-based measures of performance require generation of an equity curve and then analysis of the curve to determine the value of the statistic on which the choice of model is to be based. However, the *Profit Factor* only requires summary statistics of the trades that would have occurred if the model and trading rules had actually been applied during the modeling period.

6.7 USING KERNEL REGRESSION FOR CLASSIFICATION PROBLEMS

Classification is a well-known statistical activity[8] whose purpose is to develop criteria for making decisions regarding identification with a group. Something is measured or observed, and it must then be assigned to one of several groups. An early giant in the field of statistics, R. A. Fisher, first studied this problem in 1936 and developed what is today known as the Fisher discriminant function. Many financial modeling problems fall within the realm of classification problems.

As an example of a problem related to trading systems, consider the development of a filter for discriminating between several classes of trades. A two-class approach would be to identify each trade as either "good" or "bad." We could, however, turn this into a three-class problem by adding a middle class: "indeterminate." Trades in which the profit was within a specified range around zero would fall into this class. We could further complicate the classification by adding several additional groups, for example, "very good" and "very bad."

Any standard prediction problem can be turned into a classification problem. For example, assume that we are trying to predict the one-day change in the price of a financial instrument. For a straight prediction approach to this problem, the Y variable would be the price change (or perhaps the fractional price change). To turn this into a classification problem, we could assign different price ranges to different groups. The Y variable would then be the group into which each price change falls. The measure of performance might then be related to the fraction of cases in which the classification is correct (or perhaps almost correct).

As an example of a two-class problem, consider one-day price changes in which each day is marked as either "up" or "down." The purpose of the model is, at the end of each trading day, to make a prediction regarding the next trading day: is it likely to be an up day or a down day? A minor question is what do we do with days in the modeling database in which the change is zero. Several possible methods may be used to treat records that don't fall into either class. Typically, such records are assigned an intermediate value. However, an alternative approach is simply to discard these records.

To apply kernel regression to this two-class problem, we can assign values of $Y = 1$ to all up days and -1 to all down days. For zero-change dates (if there are any) we assign a value of 0.

Using any of the three KR algorithms (*Order 0, 1,* or *2*), we make a prediction for *Y,* and if it is positive, we interpret the prediction as indicating an up day. Conversely, negative predictions are interpreted as down days. What sort of results can we expect? In Figure 6.7 an idealized situation is considered. We assume that there is a variable X in which the values of X are distributed normally for each class. In the most general case, there would two values of μ (the mean value of each distribution) and two values of σ (the standard deviation of each distribution. In Figure 6.7 the distributions are shown as symmetric about zero with means of -1 and 1 and both with σ's of 1.

The area under each of the two curves (from $-\infty$ to ∞) is by definition one, so the area marked *Area 1* is the fraction of misclassifications that are expected for Class 1 events (for example, up days)—similarly, *Area 2* is the fraction expected for Class 2 events. For the specific case shown in Figure 6.7, our theoretical expectation is 84.1 percent correct classifications for both classes. (The integral of the standard normal distribution from $-\infty$ to 1 is 0.84.) As the two distributions approach each other, the percentage of correct classifications approaches 50 percent. For example, if the two means were moved to -0.1 and 0.1, the theoretical percentage of correct classifications would be 54.0 percent.

One measure of the *goodness* of any classification methodology is to see how closely it approaches the theoretical limit. The following experiment was performed on 10 different sets of data similar to the two classes shown in Figure 6.7 but with the mean values at -0.1 and 0.1. The theoretical correct classification for this data is 54.0 percent. For each set of data, 10,000 records were used as learning data, and the classifications were made on 5000 test data points. For this very unique data set, the best results are obtained using a tree of height 1 (i.e., only two cells). For the 10 data sets, an average value of 53.86 percent was observed with a σ of 0.71 percent. In other words, some data sets actually outperformed the theoretical value of 54.0 percent. Although the order of the data records was randomly distributed between the two classes, the average value of X was close to zero and so the X space was split at a point near zero. It is therefore not surprising that the average result for a tree-height of one was very close to the theoretical value.

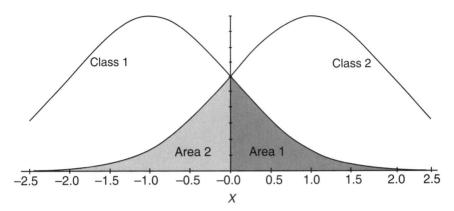

Figure 6.7   Distribution of two classes as functions of X.

For the case of one-dimensional simple models similar to the model shown in Figure 6.7, use of sophisticated techniques such as kernel regression or neural nets offers no advantage. Simple classical statistical methods are more than adequate to find an optimal solution for the classification problem. Even for higher dimensional models, classical methods are adequate if there is a simple split between the two classes. For an example, consider Figure 6.8. In this figure we see that Class 1 data points reside primarily in the upper left quadrant and Class 2 resides primarily in the lower right quadrant of the $X1$–$X2$ plane. There is, however, some overlap. If the classification is based on a split

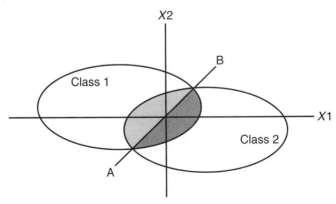

Figure 6.8   Two classes distributed as functions of $X1$ and $X2$.

using the line *AB*, records from Class 2 falling to the left of the line will be misclassified. Similarly, Class 1 records falling to the right of the line will also be misclassified. Finding the optimal line AB is simple using classical methods. Using a p-tree as described in Chapter 4, it is clear that we need a treeheight large enough that there will be enough cells to break up this space in a manner that approximates the split along line AB.

We can, of course, show examples in which a single-line discriminator in a 2D space is inadequate. Figure 6.9 illustrates this point. In this figure, the records from both classes are distributed in two overlapping regions. One line cannot possibly split this plane into two regions such that the two classes reside primarily on different sides of the line. The p-tree split as described in Chapter 4 provides a simple solution to this complicated situation. Assume that the distribution within each of the four large oval areas by themselves is constant and that where the large areas overlap, the data densities are the sum of the densities within the large areas. For example, the areas marked A would have double the density of the areas marked Class 1 and 2. The areas marked B would have triple density, and the C areas would have quadruple density.

A KR analysis of this space is initiated by dividing the space into a number of cells of approximately equal population. Cells in the Class 1 and 2 areas will be populated by homogeneous populations of the same class. Cells primarily in the A and C

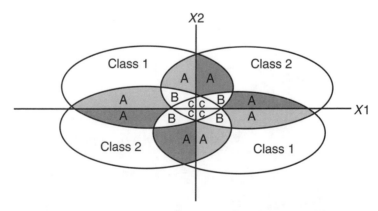

Figure 6.9 A more complicated two-class distribution.

areas will have mixed populations of approximately equal numbers of learning data points from each class. The B areas will have approximately twice as many of one class as compared to the other. For example, the B area in the lower left quadrant should have about two Class 1 records for every Class 2 record. For a cell exclusively in this area, we would thus expect classifications for any test point falling into this area to be Class 1. In other words, for test points falling into this area we would expect misclassification for about one-third of the test records.

The point to be noted when comparing Figures 6.8 and 6.9 is that we can easily move from a situation in which classical statistical methods are applicable to complex cases where we require more sophisticated tools. As the dimensionality of the model increases, the probability of finding a simple solution to the classification problem becomes less and less likely. In addition, as the number of classes increases, the need for more sophisticated methods becomes more evident.

6.8 FINE TUNING A MODEL

As shown in Section 5.5, computer time differences due to the choice of parameters can be substantial. In particular, the *fast mode* of operation (which avoids a nearest neighbor search and uses only the data points in the test cells) can lead to large time savings. Alternatively, by not using the *fast mode,* the predictability of the model can be significantly improved. One approach to modeling is to exploit both modes of operation.

If the number of data points is not large and *ncp* (the number of candidate predictors) is relatively small, then there is no need to use the *fast mode* of operation. The definition of "large" and "relatively small" varies from machine to machine. However, the user can easily determine the relative times of the different modes of operation by making a few simple tests. If one sees that the total time required is excessive if *fast mode* is not used, then a combination strategy should be considered.

For example, consider a hypothetical case in which we are interested in finding a model based on *ncp* = 100. Let us say we will look at all one and two-dimensional spaces and then use the best twenty 2D spaces to create 3D spaces and the best twenty 3D spaces to create 4D spaces. The maximum number of spaces that will be examined is:

$$Max\_Num\_Spaces = 100 + 100*99/2 + 20*98 + 20*97 = 8950$$

After a preliminary analysis of a few spaces, we see that the average time per space using the *fast mode* is 0.3 seconds and without the *fast mode*, 10 seconds. Thus a full analysis with *fast mode* will take about 2685 seconds, and without it the expected time is 89500 seconds (i.e., more than a day). Furthermore, if we wish to do a complete parameter search varying such parameters as *numnn* (the number of nearest neighbors) and K (the smoothing constant), we will have to repeat the long run for each combination of parameters.

To speed up the process, the following combination strategy can be used. First use the *fast mode* to find the top N models. Then without fast mode do a complete parameter study of these N models. If, for example, the number of combinations of parameters is M then the total time for the parameter study is $10NM$, The choices for N and M can be made based on the amount of computer time allotted to the analysis. If, for example, we want to start a run at the end of the day and look at the results the next morning, we can budget about 10 hours for the run. In other words, the parameters N and M should be set such that the product NM is approximately 3600.

NOTES

1. J.S. Armstrong, *Long Range Forecasting: From Crystal Ball to Computer,* Second Edition (New York: John Wiley & Sons, 1985).

2. J. M. Bates and C. Granger, "The Combination of Forecasts." *Operations Research Quarterly,* 20 (1969): 451–469.

3. J. E. Freund, *Mathematical Statistics* (Englewood Cliffs, NJ: Prentice Hall, 1992).

4. S. Siegel and N. J. Castellan, *Nonparametric Statistics* (New York: McGraw-Hill, 1988).

5. W. W. Daniel, *Applied Nonparametric Statistics* (Boston: PWS Publishing, 1990).

6. W. Mendenhall and T. Sincich, *Statistics for Engineering and the Sciences,* (New York: Macmillan Publishing, 1992).

7. B. B. Mandelbrot, "A Multifractal Walk Down Wall Street," *Scientific American* (Feburary 1999).

8. P. R. Krishnaiah and L. N. Kanal, Classification, Pattern Recognition and Reduction of Dimensionality, Volume 2, *Handbook of Statistics* (North Holland, 1982).

7

CREATING TRADING SYSTEMS

7.1 TRADING SYSTEMS

One of the primary purposes of modeling financial markets is to develop trading systems. If the model has real predictive power, then to exploit the model one needs an approach to using it in a trading system. There is a large body of literature on trading systems. A recent book by Kaufman covers a whole range of subjects, including trading systems, approaches, and indicators.[1] Systems and indicators are described in sufficient detail for programmers to turn the concepts into software. Schwager's book is a tour-de-force on the application of the full range of technical analysis.[2] It covers chart patterns and interpretation, indicators, trading systems, and system testing. A book by Williams provides a number of specific trading patterns for short-term swing trading and for complete trading systems.[3] The interesting use of certain calendar and seasonal effects to filter signals is discussed. A book by Pardo discusses the formulation and testing of trading systems.[4] A general approach is described that can be used regardless of the type of system being tested. O'Shaughnessy's book is quantitative and discusses which variables can forecast superior and inferior stock returns using a 40-year database.[5] A book by Ruggiero covers a range of traditional and advanced data analysis and modeling methods and their use in developing computer-based trading strategies.[6] A book edited by Jurik includes several interesting case studies that utilize advanced modeling methods.[7] Chande and Kroll offer some useful ideas regarding the development of technical inputs to computer models,[8] and a more recent book by Chandee includes an extensive bibliography on the entire subject of trading systems.[9]

Most of the well-known trading systems are rule-based and use only the price data of the financial instrument being traded. Rule-based systems have an obvious appeal: the user can understand why he or she is initiating a buy or sell order. In addition, the system usually includes rules for exiting a trade. Thus the system can be completely mechanical. For example consider a simple *channel breakout* system. The rules are quite simple:

1. If today's closing price is higher than the high over the last N days, buy on the close.

2. If today's closing price is lower than the low over the last N days, sell on the close.

3. Close out the long trades at the low over the last M days (where M is less than N).

4. Close out the short trades at the high over the last M days.

Many simple systems such as this one are in widespread use today. Using readily available products such as Omega Research's *TradeStation*™, one can easily implement these systems on a home computer and use them to enter and close trades. However, these simple systems based on only price data do not exploit the true potential of the computer hardware readily available today.

There are several different approaches to exploiting rapid multidimensional methods such as kernel regression in the development of trading systems. We can use KR to develop a model and then build a trading system around the model. An alternative approach is to use an existing trading system (for example, a simple rule-based system) and then use KR to improve the system through a filtering mechanism.[10] What is gained by these approaches is that information available in relevant series can be used to improve the performance of trading systems. For example, it might be possible to discover interest rate effects on price changes for a particular class of financial instruments.

A major problem involved in developing trading systems is the amount of available data. For example, consider the development of a filter for a simple rule-based trading system that uses daily price data. If 10 years of data are available, then there will be about 2500 days of data. If we are looking at a single financial instrument, then we will have only 2500 data records. If the trading system initiates on the average one trade per month, the total number of trades in the database will be about 120. When one considers the number of parameters that can be varied, it seems quite obvious that we can find a set of parameters that will show excellent performance for this small number of trades. For example, consider the channel breakout system described above. By varying M and N we can get a large spread in performance from highly profitable to catastrophic. By adding slightly more complex entry and exit criteria, we can considerably expand the performance range. The only question remaining is: Will the system hold up out-of-sample?

If one looks carefully at the best performance of trading systems that are based on a relatively small number of trades in the modeling period, the reason for the success of the system is usually quite obvious. The optimum parameters were determined such that a few major moves were entered near the bottom and closed near the top. There is absolutely no guarantee that this performance will be repeated in an out-of-sample period or in actual usage. One can ask the question: Can we improve the system by filtering with KR or a similar technique? Unfortunately, if the basic trading system was developed from the results of a relatively small number of trades, sophisticated methods of analysis will not correct this basic flaw in the methodology. Simply stated, one cannot develop robust trading systems based on the results of a relatively small number of trades in the modeling period.

So what can be done? The only answer that makes sense is to increase the number of simulated trades to a "large" number. This can be done in several ways. The most obvious method for increasing the number of trades is to include a number of similar financial instruments in the trading database. For example, rather than trying to develop a system for a single common stock, by considering a group of stocks the number of trades can be increased roughly proportionately to the number of companies included in the group. Another possibility is to use intraday data to develop a much faster trading system. If, for example, the average number of trades per day is two, then 10 years of data will result in about 5000 trades. This number at least offers the hope of obtaining meaningful results.

A different approach to trading is to use computer methods for portfolio selection. For example, KR can be useful in stock selection. For such applications the purpose of the model is to make predictions as to which stocks will show the greatest increases in value and which will show the greatest decreases. Using a market neutral strategy, one can then choose two groups of stocks. Long positions will be taken for those stocks in which the predicted returns are the greatest, and short positions will be taken for those with the worst predicted returns.

7.2 GENERATING SIGNALS

The most straightforward use of prediction models is the generation of signals for trading systems. If a model has passed testing and exhibits predictive power, the next step is to use the real-time predictions for trading. The task facing the analyst is to develop a system for turning predictions into trading signals. A typical approach to this problem is to determine several parameters that specify trading rules. For example:

1. If the current position is neutral and the value of the prediction goes above a certain level, enter a trade on the long side. (We can call this value the *buy_entry_threshold.*)

2. If the current position is neutral and the value of the prediction goes below a certain level, enter a trade on the short side. (We can call this value the *sell_entry_threshold*).

3. If the current position is long and the value of the prediction goes below a certain level, then close the trade. (We can call this value the *buy_cover_point*).

4. If the current position is short and the value of the prediction goes above a certain level, then close the trade. (We can call this value the *sell_cover_point*.)

This simple scheme has the attractive property that there are only four free parameters that specify the trading rules. Once again, it should be emphasized that the greater the number of free parameters used to specify any system, the greater the probability of overfitting.

To generalize the scheme, the four parameters can be made dimensionless if they are specified as fractions of the maximum values obtained in the model evaluation period. An alternative procedure is to sort the predictions in the evaluation period and set the parameters on the basis of percentiles of the sorted predictions. An idealized example is shown in Figure 7.1. The X-axis represents the normalized predictions. A value of 1 is assigned to the maximum positive prediction and −1 to the maximum negative prediction. A value of zero is assigned to predictions of zero. In other words, the scaling of negative and positive predictions might not be the same. The Y-axis is the percentile of the distribution from −1 to X. If, for example, we set the *buy_entry_threshold* to correspond to the eightieth percentile,

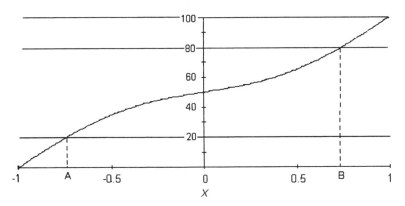

Figure 7.1 Cummulative probability distribution of the predictions.

then the entry point for long positions will correspond to point B. Similarly, if the *sell_entry_threshold* is set to correspond to the twentieth percentile, then the entry point for short positions will correspond to point A.

For this idealized example, since the cumulative probability curve crosses the Y axis at 50, 50 percent of the predictions are negative and 50 percent positive. In reality, however, there is no reason to expect such an outcome. If the percentage of positive predictions is greater than 50 percent, then the curve will cross the Y axis at a value less than 50. If the average change was positive during the modeling period for the market (or markets) being modeled, then we would expect the fraction of positive predictions to be greater than 0.5. Similarly, if the average change were negative, the expected fraction would be less than 0.5.

7.3 CREATING A FILTER

Filters can be used to exploit information from a variety of sources to improve the performance of trading systems. The concept of using filters was first proposed by Aronson in 1991.[10] Typically, the starting point is a simple rule-based trading system such as the *channel breakout system* described in Section 7.1. The function of a filter is to treat the trades generated by the trading system as a *classification* problem. The initial trading system generates entry and exit signals. The purpose of the filter is to make the final decision: do we accept or reject a signal issued by the trading system?

Once again, the requirement for sufficient data mentioned in Section 7.1 should be emphasized. If the number of simulated trades issued during the modeling period is small, then there is no hope of creating a robust filter that will add value in real time. However, if there are a "reasonable" number of simulated trades in the modeling database, filtering can be a useful activity. To appreciate the concept, consider typical results shown in Tables 7.1 and 7.2.

The *profit factor* is defined as the ratio of the profit on winning trades divided by the loss on losing trades:

TABLE 7.1 Trade Simulation Results Without Filtering

	Short Trades	Long Trades	All Trades
Number of trades	743	921	1664
Percent winners	44.1	46.9	45.7
Profit factor	0.871	0.949	0.909

$$profit\_factor = \frac{num\_winners * avg\_prof\_per\_win\_trade}{num\_losers * avg\_loss\_per\_losing\_trade} \quad (7.1)$$

A value of *profit factor* less than one implies that the trading system will lose money over the long run. A value greater than one implies a profitable trading system. The results from Figure 7.1 are not encouraging. In Figure 7.2 we see the results after filtering. The number of trades has been reduced by approximately one half. As a result of filtering, the combined system appears to be profitable as opposed to a losing system prior to filtering. Although there is no guarantee that the profitability of the filtered system will continue in real time, the results are encouraging and the system clearly might lead to real profits in the future.

These results do not provide the full story. For example, equity-curve statistics such as return-on-investment, the Sharpe Ratio, and drawdown details are missing. To obtain these and other equity curve statistics, one would have to generate equity curves and then use some sort of trade simulator to obtain the detailed results. For example, the summary tables from an output report of RTSIM™ (a product of Raden Research Group and Riskon) are included in Figures 7.2 and 7.3. These tables are printed with the permission of Innervision Asset

TABLE 7.2 Trade Simulation Results with Filtering

	Short Trades	Long Trades	All Trades
Number of trades	360	410	770
Percent winners	51.9	48.7	50.2
Profit factor	1.818	1.367	1.575

```
Aggregate Portfolio Trading Statistics from 19850101 to 19980608

          Round   Profit  Percent  Total    Total    Comm /  Turns /
   Asset  Turns   Factor    Win     Comm      Slip    FinEq   Mil-yr
    Port  79462    3.49    61.2    114407   732978   0.009   2969.89
 1   SP    2941   10.32    73.6    294100        0   0.024    219.84
 2   UB    3361    4.77    67.6     33725   488205   0.006    251.24
 3   ED   11035    2.88    57.5    110350   868600   0.027    824.87
 4   OL    6650    2.70    60.5     73370   952000   0.019    497.09
 5   HO    7777    1.77    54.3     85663  1158885   0.028    581.33
 6   HU    5620    2.05    53.9     61980   923475   0.022    422.87
 7   JY    3222    3.42    64.2     32360   289375   0.007    240.85
 8   BP    2239    4.82    59.7     22455   108719   0.007    167.37
 9   DM    3909    3.05    62.6     39160   383500   0.009    292.20
10   GB    3742    2.56    61.0     36494   188059   0.013    398.54
11   NN    5003    4.32    62.3     58001   146101   0.015    423.93
12   LG    2518    2.98    59.4     26991   491952   0.008    188.22
13   IB    1471    2.41    53.2     17118   150231   0.008    220.86
14   FT    4520    7.34    69.7     49793    61628   0.004    397.22
15   MX    2330    2.42    57.6     31358    32869   0.010    241.67
16   DX     762    4.15    69.6     10571    22389   0.004    103.02
17   GC    5873    3.14    69.1     64719   232825   0.020    439.01
18   SI    4584    2.02    57.2     50551   766500   0.022    342.66
19   BT     440    2.87    54.8     28679    62369   0.009     54.02
20   NX    1465    2.48    57.9     16632     2103   0.006    131.49
```

Figure 7.2 Trading statistics report from the RTSIM™ output report.

```
Summary Portfolio Statistics from 19850101 to 19980608

         Average     Max      Avg    Advocom   Sharpe  MAR Sharpe
   Asset    ROI    Drawdown  Drawdown  Ratio    Ratio    Ratio
    Port   39.91     6.92      4.46    32.56     3.25     1.98
 1   SP    84.80    31.94     12.95    12.31     1.85     1.29
 2   UB    32.90    34.89     12.37     3.74     1.84     1.70
 3   ED    22.90    62.73     20.58     1.35     0.72     1.04
 4   OL    21.56    45.78     16.39     1.78     1.05     0.73
 5   HO    15.59    83.84     22.06     0.69     0.54     0.80
 6   HU    13.38    40.61     14.57     1.35     0.71     0.81
 7   JY    27.18    25.70     11.65     3.94     1.38     1.27
 8   BP    17.21    11.40      6.55     4.96     1.36     1.34
 9   DM    23.71    29.89     11.81     2.83     1.42     1.15
10   GB    20.08    23.03     15.20     2.39     1.13     1.09
11   NN    23.64    50.91     14.48     4.37     1.02     1.62
12   LG    18.36    91.84     28.22     1.06     0.42     0.39
13   IB    19.16    16.27      9.14     4.16     1.19     1.18
14   FT   100.84    33.85     11.66    13.41     1.86     1.07
15   MX    20.81    18.88      9.21     4.00     1.22     1.07
16   DX    25.29    13.73      7.29     7.30     1.41     1.75
17   GC    16.92    23.99      7.76     3.08     1.17     1.28
18   SI     9.43    24.81     10.22     1.24     0.75     0.73
19   BT    25.16    21.16     11.15     4.01     1.22     1.44
20   NX    17.16    17.65      9.79     3.12     1.06     1.07
```

Figure 7.3 Equity statistics report from the RTSIM™ output report.

Management and are the results of a hypothetical simulation of a trading system using data from 1985 through June 1998. (These results do not represent trading results from any account that Innervision Asset Management has actually traded.) The system for trading futures contracts was applied to 20 different asset types. (For example, SP is their code for S&P Futures.) The total number of simulated trades is almost 80000, which is more than sufficient to obtain reasonable statistics.

An interesting point to note in these results is the effect of simulating an entire portfolio as opposed to a single asset. From the equity curve statistics we see that the largest drawdown in the modeling period was 6.92 percent, whereas the drawdowns for the individual assets were much greater. For one asset (LG), the maximum drawdown was almost 92 percent! The simulated Rate-of-Return for the portfolio was close to 40 percent per year for the modeling period, with a Sharpe Ratio of 3.25.

7.4 CROSS-SECTIONAL TRADING SYSTEMS

Trading systems related to a single financial instrument are almost always based on intraday data. Using tick data or bar data in which each bar represents a few minutes of trading, sufficient data can be obtained to make modeling of single financial instruments feasible. Typically, the trading systems using such models are for intraday trading, and the average number of trades per day can be more than one. However, for trading systems in which the time horizons are much longer (i.e., days, weeks, or months), it is extremely difficult to model a single financial instrument. The reason for this difficulty is the lack of sufficient data.

A reasonable question is: why can't we use bar data (let's say five-minute bars) to develop models with longer time horizons (e.g., several days)? Clearly, some relationship must exist between the time scales of the candidate predictors and the time scale of the predictions. For example, if we wanted to make a weather prediction for three days into the future, the weather trend over the last hour would hardly provide a reasonable database for making the prediction. Similarly, the weather trend over the last week would not be useful in making a prediction about the change in the weather over the next hour. Financial markets are quite similar. If we wish to make a prediction about the change in the daily or weekly price of some financial instrument, the candidate predictors must have compatible time scales.

As an example, consider a proposed trading system in which the objective is to look at closing prices at the end of a trading day and then make decisions regarding the next day. Such a system is based on a daily time scale, and we will be forced to use a database of daily data. Some candidate predictors might include summary statistics regarding price movements within the day, but these indicators would be daily in nature. An example of such an indicator is the ratio (*close-open*) / (*high-low*). In addition, we would also require candidate predictors in which the trends over several days would be captured. The amount of daily records is limited to about 250 per year. To overcome this limitation, as explained in Section 6.4, the obvious solution is to use cross-sectional data. Designing trading systems that extend over several financial instruments is, however, more complicated than designing trading systems for a single instrument.

A major problem associated with cross-sectional trading systems is money management. In Figures 7.2 and 7.3, simulated results are included for a 20-asset futures portfolio. The results are taken from an output report of the RTSIM™ program (a product of Raden Research Group and Riskon). We see that, although the overall performance is relatively stable, the equity swings in the individual assets are dramatic. This fact can be verified by observing the values of maximum drawdown. (*Drawdown* is defined as the maximum dip in equity from the last previous high. In Figure 7.3, it is presented as a percentage.) Although the maximum drawdown for the entire simulated portfolio was 6.9 percent, a value of almost 92 percent was observed for asset LG. When some assets have significantly increased in value and others have lost significant value, what should be done? Redistribution of funds among the various assets is one of the money management issues that must be included as part of the trading system design.

There are several hidden layers of complexity between the model and simulated or actual results. To obtain results similar to those shown in Figures 7.2 and 7.3, a file of signals must be generated. The technique for generating signals is different if the model is a pure prediction model or is based on filtering a technical system. For models based on filtering, a signal generator based on the technical system is first used to create a set of unfiltered signals. The filter model is then used to determine how to react to each signal. For prediction models, logic for creating signals from the predictions is required.

Another problem is scaling. As the size of the portfolio grows or shrinks, the actual size of the orders may be adjusted. For example, if we are simulating a stock portfolio, the number of shares bought or sold when a signal is issued is a parameter that might vary as part of the simulation. For a futures portfolio, the number of contracts per signal might be variable.

7.5 UPDATING MODELS

In Section 6.3 the modeling of dynamic systems was discussed. The nature of financial markets is dynamic, and a model that has been useful for a period of time will eventually become outdated. In other words, profitable models can be expected to become unprofitable sometime in the future. Hopefully, from simulation, the analyst can get a feel for a "reasonable lifetime" for a model. What should be built into any trading system is a procedure for periodically updating models.

The problem with a term such as *reasonable lifetime* is that it is hard to define. Even the best models can go through bad periods. In Section 2.5 Eq. (2.13) was used to predict the probability of a drawdown:

$$\text{Prob(Drawdown} >= P) = \left(1 - P/100\right)^{2\mu/\sigma^2} \tag{2.13}$$

This equation is based on a very simple model: the equity changes are normally distributed with a mean of μ and a standard deviation of σ. In this equation, μ is the average fractional change in equity, and σ is the standard deviation of the equity changes. If we are using daily data, the equation predicts the fraction of days in which the drawdown is greater or equal to P. If we consider longer time periods, the same equation is valid and the value of the exponent remains essentially the same. For example, if there are 250 trading days per year, the number of days per month is approximately 250/12. The monthly value of μ would therefore be:

$$\mu_{monthly} = (1 + \mu_{daily})^{days\_per\_month} - 1 \qquad (7.2)$$

For small values of μ_{daily} this equation reduces to approximately:

$$\mu_{monthly} \cong days\_per\_month * \mu_{daily} \qquad (7.3)$$

The value of σ^2 scales as $days\_per\_month$, so the exponent in Eq. (2.13) is approximately the same for daily and monthly data. In fact, it is approximately independent of the time scale.

Equation (2.13) can be used to provide a quantitative method for testing whether or not a trading system has "broken down." If during the simulation period the trading system exhibits a fractional monthly return of μ with a standard deviation of σ, the drawdown probability can be used to test the validity of the system. Using, for example, a 5 percent level of confidence, we can compute the size of the drawdown level that should be exceeded by chance for 5 percent of the months. If the drawdown does in fact exceed this limit, then we might arbitrarily decide that it is time to update the model. As an example, consider a system in which the simulated results show an annual return of 30 percent with a daily ratio of $\sigma/\mu = 10$. Using 250 trading days per year, the daily value of μ is $1.3^{-250} - 1 = 0.00105$. The value of σ is therefore 0.0105, and the exponent in Eq. (2.13) is 19.05. The value of $1 - P/100$ computed from this equation is $0.05^{-19.05} = 0.854$. In other words, the probability of a drawdown of at least 14.6 percent can be expected for 5 percent of the months. Note that the definition of drawdown is based on the previous high, so a drop of 14.6 percent does not imply a monthly drop of this magnitude. It implies a cumulative drop from the previous high. This is an arbitrary criterion, but it does provide a simple method for making a decision.

Another approach to updating the models is to allow models to run during the simulation period and see how long it takes to obtain a specified level of drawdown. By altering the modeling period, a number of different models and drawdown statistics can be obtained. Analysis of the results might yield an insight into the "expected lifetime" of models obtained in this fashion. This expected lifetime could then be used in real time. For example, consider a situation in which most of the models are profitable during the first six months beyond the modeling period, and then after about a year the results seem increasingly erratic and unprofitable. It should be clear that the choice for the expected lifetime should be about six months.

For institutional trading, the cost of modeling is negligible compared to the potential trading profit or loss. For this reason, one might conclude that the modeling process should be continual. For example, if we are using daily data, why not repeat the entire modeling process at the close of each trading day? If the modeling database extends back in time for several years, moving the data window forward by one day should not result in a serious change in the model. If the analyst decides to remodel on a regular fixed schedule, one approach is to decide the fractional level of replacement of the database. For example, if the model is based on 500 trading days and we set the replacement level at 10 percent, we will move the database window forward by 50 days after every 50 trading days. Dropping 10 percent of the earliest data and replacing them with 50 new days of data should make a noticeable difference in the resulting model. Even if the *growing* or *moving* option is used, remodeling might result in an entirely different selection of variables. The choice of 10 percent is arbitrary. One can determine an "optimum" value through the simulation process.

7.6 STOCK SELECTION: A CASE STUDY

Sections 6.3 and 6.4 made reference to a stock selection trading system. In this section, this system is considered as a case study. Emphasis is placed on the modeling aspects of the project. Kernel regression software was used to develop models based on a large database of monthly stock data.

The project was initiated to test the concept of combining available indicators to create added value. There is a wide body of literature related to the generation of indicators that are useful for forecasting relative returns of stocks.[5] Indicators are typically generated using stock market prices and performance as well as income statement and balance sheet statistics. Several companies provide indicators on a monthly basis. We used 11 indicators provided by the BARRA Corporation for a large body of stocks over a 15-year period. The indicators were first shown to have predictive power on their own. The open question was: Could they be combined in such a manner as to improve the predictive power in a significant manner?

The amount of data available was large: the database included monthly data for thousands of different common stocks going back to 1984. However, not all stocks were included in the database for the entire period. Stocks were continually being added and removed from the database, but the overall trend was an increasing number over time. For a given stock and a given month, a measure of performance was required as the modeling criterion. The chosen measure was the excess return for the next month. The excess return is defined as the actual percentage return minus the average percentage return for all stocks in the database for the same month.

The modeling database included 15 columns of data: the 11 indicators, the actual return for the next month, the excess return for the next month, a column containing the stock code, and a date column indicating the month. The actual returns were not used in the modeling process but were included to allow cross checking of the data. The raw data included some missing values for most of the stocks. Only records with all 11 indicators available for a given stock and month were used. A program was written to convert the raw data files into a single modeling database. After eliminating records with any missing values, the total number of remaining records in the modeling period (January 1984 to December 1998) was about 291,000 over 180 months. The average number of stocks included for each month over the entire period was therefore about 1620 (i.e., 291619/180), but the actual number on a monthly basis started at less than 1000 and built up to over 2000. The modeling criterion used in the analyses was SEP (i.e., group separation as defined by Eq. 6.7). A value of 10 percent for the group percentages was used to define the top and bottom groups. The top group is the group in which the predicted values of excess return are the greatest; the bottom group is the group in which the predicted values of excess return are the

```
Output report:   FKR Analysis Sun Jun 06 13:12:44 1999
                 Version 3.12  June 6, 1999

Parameters used in analysis    PARAMETER FILE: barra.fpa
                               INPUT DATA FILE: barra.pri

DMIN    - min number of dimensions            :      2
DMAX    - max number of dimensions            :      3
NCP     - number of candidate predictors      :     11
NCOL    - number of columns of data           :     15
NREC    - total number of records             :291619
NFOLDS  - number of evaluation folds          :     14
LTYPE   - Learning set type (G, M or S)       :      G
          G is growing, M is moving and S is static
GROWTH_FACTOR - (max_num/leaf)/(num/leaf)     : 10.00
NUMCELL- Number of cells searched for NUMNN   :      1
NUMNN   - Irrelevant because FAST mode specified
STARTVAR - starting variable                  :      1
DELTA   - Minimum incremental change    :  -100.00
CLIP    - Clipping parm for Y values    :    50.00
NODATA  - For Ycalc (columns of YCF file):   -999.00

Kernel Regression parameters:
MOD_CRITERION- 1=VR, 2=CC, 3=CC_O, 4=Separation :     4
GROUP_PCNT    - Group Parameter (for Separation) :  10.0
FIT_ORDER - 0 is average,  1 & 2 are surfaces   :     0
   FAST option used: Ycalc is cell average
NUMK - Number of smoothing parameters         :      1
K(1) - Smoothing parameter 1                  :   1.00
YCOL - Column of dependent variable           :     13

Tree parameters:
BUCKETSIZE - design number per cell (computed) :  1007
TREEHEIGHT - tree parm                         :     6
Computed number of cells                       :   127
Computed number of leaf cells                  :    64
Computed avg bucket size                       :1007.8
Computed 2 Sigma limit for F Stat              :  1.36
```

Figure 7.4 Parameter settings for FKR stock selection run.

least. Parameter settings for a typical FKR analysis are included in Figure 7.4. This run was based on the *fast mode* of operation with *Order 0* and *treeheight* = 6. Each space was therefore divided into 64 (i.e., 2^6) cells. Another parameter of interest is the *clip* parameter. This parameter was used to limit outliers (i.e., extreme values of excess return) to the specified range: –50 to 50. Results for fold 1 are shown in Figure 7.5. A total of 14 folds were included in this run, and each fold included six months of out-of-sample evaluation data. The results varied from fold to fold and are summarized in Table 7.3.

If the predicted and actual values of excess return are unrelated (i.e., the model is useless), SEP would be expected to be distributed normally with a mean of zero and a standard deviation of 1. In other words, SEP measures the separation between the means of the two groups in units of standard deviations. The SEP values included in Table 7.3, with the exception of fold

```
Data set parameters for fold 1 of 14 folds:
□
  NLRNDATES - number of learning dates      :    24
  NTSTDATES - number of testing dates       :    12
  NEVLDATES - number of evaluate dates      :     6
  GAPDATES - gap between data sets          :     0
  STARTLRNDATE - start learning date        :  8901
  STARTTSTDATE - start testing date         :  9101
  STARTEVLDATE - start evaluate date        :  9201
  NLRN    - number of learning records      : 34265
  NTST    - number of test records          : 17629
  NEVL    - number of evaluation records    :  9073
  STARTLRN - starting learning record       : 75262
  STARTTST - starting test record           :109527
  STARTEVL - starting evaluation record     :127156

Ordered results for 2 dimensional models:
    Total number of combinations:      55
    Number tested             :        55
    survivenum( 2)            :         5
    Number of survivors       :         5
Sep:    6.422   FracSS: 0.552   F:   5.87   X:  5   11
Sep:    6.051   FracSS: 0.554   F:   3.72   X:  2    9
Sep:    5.896   FracSS: 0.553   F:   7.19   X:  9   11
Sep:    5.895   FracSS: 0.533   F:   5.21   X: 10   11
Sep:    5.659   FracSS: 0.542   F:   5.92   X:  4   11
```

Figure 7.5 Fold 1 parameters and results for 2D spaces (part 1).

```
Ordered results for 3 dimensional models:
☐
   Total number of combinations:     165
   Number tested            :      38
   survivenum( 3)           :       5
   Number of survivors      :       5
Sep:   6.896   FracSS: 0.549   F:  5.77   X:  5  10  11
Sep:   6.890   FracSS: 0.546   F:  5.05   X:  6  10  11
Sep:   6.593   FracSS: 0.551   F:  5.44   X:  3   5  11
Sep:   6.573   FracSS: 0.548   F:  5.57   X:  1  10  11
Sep:   6.537   FracSS: 0.549   F:  6.94   X:  9  10  11


Best Model Report Fold 1 (Test Set Data):
Model: 1 Sep:  6.896   FracSS: 0.549   F: 5.77   X:  5  10  11
Model: 2 Sep:  6.890   FracSS: 0.546   F: 5.05   X:  6  10  11


Timing Report:
   dim: 2  Num_spaces:   55   Time:  22   Time/space:  0.40
   dim: 3  Num_spaces:   38   Time:  17   Time/space:  0.45
   Total:  Num_spaces:   93   Time:  39   Time/space:  0.42


Evaluation Report Fold 1: from 9201 to 9206 (Evaluation Data)
Model: 1   Dim: 3   Sep:     4.687   FracSS: 0.542
Model: 2   Dim: 3   Sep:     3.346   FracSS: 0.544
```

Figure 7.5 Fold 1 results for 3D spaces and evaluation report num_best_models = 2 (part 2).

TABLE 7.3 Results for Out-of-Sample Evaluation Data (*fast mode, order 0, h = 6*)

Fold	Start Date	Space	SEP	FracSS
1	Jan. 1992	$X5, X10, X11$	4.69	0.542
2	July 1992	$X6, X9, X11$	7.95	0.580
3	Jan. 1993	$X5, X9, X11$	4.52	0.557
4	July 1993	$X9, X11$	2.97	0.538
5	Jan. 1994	$X4, X5, X9$	2.14	0.505
6	July 1994	$X5, X11$	4.71	0.540
7	Jan. 1995	$X10, X11$	4.73	0.556
8	July 1995	$X4, X7, X10$	2.35	0.532
9	Jan. 1996	$X2, X9, X11$	1.98	0.509
10	July 1996	$X9, X11$	0.24	0.505
11	Jan. 1997	$X5, X7, X10$	8.16	0.575
12	July 1997	$X4, X10, X11$	2.12	0.511
13	Jan. 1998	$X4, X8, X10$	4.00	0.535
14	July 1998	$X3, X9, X10$	4.38	0.534

10, are therefore quite significant. (The probability of a value greater than 1.96 is 2.5 percent for this distribution. All values of SEP except for fold 10 meet this level of confidence.) The values of *FracSS* are a measure of the fraction of the records in both groups in which the predicted and actual excess return have the same sign. For useless models, one would expect the values of *FracSS* to be distributed about a value of 0.50. The values in Table 7.3 are all above 0.50, adding further evidence that the models yield useful information.

The SEP modeling criterion indicates how well the top and bottom groups actually separate. However, when using this modeling strategy in a trading system, a measure that is closer to actual trading results is desirable. A measure that satisfies this criterion is the average excess return difference between the top and bottom groups. For any month, the difference is the expected return that would be achieved for a portfolio based on the model but neglecting management and trading costs. This return should be achieved regardless of whether the market rises or falls for the month. In Table 7.4 results are included for a portfolio in which long positions are taken for the top 10 percent of the stocks (based on the predicted excess return) and short positions are taken for the bottom 10 percent of the stocks. The results in this table were generated from the predicted and actual values of excess return included in an output file from the FKR program. The monthly return is assumed to be the difference between the realized excess return for the top

TABLE 7.4 Simulation Results for January 1987 to December 1998 Using *Fast Mode* **of Operation. Note that RR is Monthly Rate of Return**

	order 0, dim 1	order 0, dim 2&3	order 1, dim 1	order 1, dim 2&3
Average RR	1.226	1.662	1.073	1.637
Sigma RR	2.361	2.237	2.822	2.417
Avg/Sigma	0.519	0.743	0.380	0.677
Annual Sharpe Ratio	1.870	2.746	1.350	2.494
% Losing Months	23.6	18.8	36.1	20.1
Max Drawdown	8.02	6.83	7.98	7.31
Avg Drawdown	1.23	0.47	1.64	0.70

and bottom groups for the following month. For example, for the first month in the fold shown in Figure 7.4 (i.e., January 1992), the top group had an average excess return of 3.04 percent for the following month, and the bottom group had an average excess return of −2.42 percent. (Note that since short positions are held for the bottom group, a negative excess return represents a profit for this group.) The net rate of return for this month was therefore the difference: 5.46 percent. Results for simulations using all the evaluation data from January 1987 through December 1998 (i.e., 144 months) are shown in Table 7.4. The simulations include both *Order 0* and *1* and one-dimensional and multidimensional (i.e., 2D and 3D) runs.

The results in Table 7.4 include the Annual Sharpe Ratio. This ratio is a measure of the return to risk and is computed from the monthly average and standard deviation as follows:

$$Sharpe\_Ratio = ((1 + AverageRR / 100)^{12} - 1)/$$
$$((1 + SigmaRR/100)^{sqrt(12)} - 1) \qquad (7.4)$$

As Table 7.4 shows, the results using *order* 0 are marginally better than those using *order* 1. Even if one considers the costs of actually running this portfolio of several hundred stocks on both the long and short side, the results are quite impressive. Just using the best single indicator for each six-month period (i.e., *dim* = 1), we see that the average gross monthly rate of return is a respectable 1.226 percent (which is almost 16 percent on an annual basis) with an average drawdown (from the previous high) of only 1.23 percent. The maximum drawdown over the entire 12 year period was only 8 percent. If one extends the analysis to two and three dimensional spaces, the results are significantly better: an average return of 1.66 percent per month with a maximum drawdown of only 6.8 percent over 12 years.

Several similar *fast* runs with different values of *treeheight* and *order* were completed. The effect of treeheight was small but a value of 6 yielded slightly better results than the other values tested. As seen in Table 7.4, the results for *order* 0 were slightly better than those for *order* 1. The initial modeling effort was performed using the *fast* mode of operation. The reason for this choice was simply a matter of computer time. For example, the most time consuming run in Table 7.4 was the 2D and 3D analysis using *order* 1. This run took 3112 seconds to examine 2233 spaces (i.e., an average of 93 spaces per fold with average time per space of 1.4 seconds). The hardware used for this run was a Pentium P-II-400 with 128 Meg of RAM. On the other hand, running a full nearest neighbor search for each test point, an average of almost 100 seconds per space was required. To reduce the number of spaces examined using full nearest neighbor searches, the number of input parameters was reduced from 11 to 5. The five best input parameters were chosen on the basis of performance comparisons using the *fast* mode results.

The results using a full nearest neighbor search were significantly better than the *fast* mode results. Using a value of *numnn* (number of nearest neighbors) = 500, limiting the search to the test cell plus the three nearest adjacent cells (i.e., *numcells* = 4) and equally weighting each point (i.e., $K = 1$), simulation results are included in Table 7.5. These results are for simulations using 24 months of learning data, 12 months of test data and then managing a portfolio based upon the best model for the next 6 months. Each simulation required a total of 24 folds starting with the first fold from January 1987 to June 1987 and ending with the last fold from July 1998 to December 1998. The portfolio was changed monthly and the model was changed every six months. Once again, the results for *order* 0 are marginally better than those for *order* 1. The results for *order* 0 indicate an annualized gross rate of return of about 27

TABLE 7.5 Simulation Results for January 1987 to December 1998 Using Nearest Neighbor Search. *Note* that RR is Monthly Rate of Return

	order 0, dim 1	order 0, dim 2&3	order 1, dim 1	order 1, dim 2&3
Average RR	1.542	2.025	1.636	1.924
Sigma RR	2.711	2.314	2.863	2.369
Avg/Sigma	0.569	0.875	0.571	0.812
Annual Sharpe Ratio	2.076	3.297	2.093	3.041
% Losing Months	24.3	16.0	25.0	17.4
Max Drawdown	8.89	8.05	8.45	8.40
Avg Drawdown	1.06	0.33	1.33	0.43

percent over the 12 year simulation period. The maximum drawdowns of 8.05 percent (for *order* 0) and 8.4 percent (for *order* 1) occurred entirely in one terrible month of trading: October 1996. For *order* 0, the "worst" group showed an average increase in value of 6.3% (i.e., a loss of 6.3% because short positions were held for these stocks) while the "best" group lost an average of 1.75%. This particular month illustrates the need for building some sort of *stop-loss* strategy into trading systems.

A concept discussed in Section 6.5 is combining models. In Tables 7.4 and 7.5 results were obtained using only the best model predictions in each fold. In Table 7.6 the results of using predictions that are the average of several models are included. The results for the best single model is seen in Table 7.5 for *order* 0, *dim 2 & 3*. We see in Table 7.6 that for this particular problem there is a substantial performance improvement achieved by using average predictions as compared to just the single best model predictions. For example, in Table 7.5 we see that the average monthly rate of return for the best model over the 144 month period was 2.025 percent. Using the average of the best five models increases the monthly average to 2.609 percent (i.e., 36 percent on an annual basis). The Sharpe ratio increases from 3.297 to 4.159, and the maximum drawdown decreases from 8.05 percent to 5.91 percent.

The next phase of the modeling process focused on the expansion of the candidate predictor space. Besides the 11

TABLE 7.6 Simulation Results for January 1987 to December 1998 Using Average Predictions from Several Models. All Results Are Based Upon *order* 0, *dim 2&3*. Note that RR is Monthly Rate of Return

	top 2 best models	top 3 best models	top 4 best models	top 5 best models
Average RR	2.274	2.351	2.531	2.609
Sigma RR	2.436	2.441	2.443	2.439
Avg/Sigma	0.933	0.963	1.036	1.070
Annual Sharpe Ratio	3.562	3.691	4.011	4.159
% Losing Months	15.28	15.97	11.11	11.81
Max Drawdown	7.68	7.74	7.49	5.91
Avg Drawdown	0.36	0.29	0.22	0.21

original predictors, the 1 month, 2 month and 4 month differences of each of the 11 predictors were included in an expanded database. In other words, the number of candidate predictors was expanded from 11 to 44. Only spaces with at least one of the original 11 candidate predictors were examined. (The concept of forcing particular predictors into subspaces is discussed in Section 4.7.) The number of 2D spaces considered included all combinations of the original 11 predictors (i.e., 11*10/2 = 55) plus all pairs made between these 11 and the new 33 predictors for a total of 418 2D spaces (i.e., 55 + 11*33). The analysis was continued to the 3D level using the best 5 spaces as survivors of the 2D analysis. Unfortunately, the preliminary results using the *fast* mode of operation were not as good as the results achieved using just the original 11 indicators, so this line of inquiry was not pursued any further.

It should be emphasized that the modeling strategy developed in this section is for a large system of the type that might be considered by hedge funds. It is clearly not applicable for a single investor. Basing a trading system upon this modeling strategy requires coordinating a monthly reevaluation of a large portfolio of stocks. Developing a computerized system to turn the modeling strategy into a trading system is a nontrivial matter and is not within the scope of this book. However, some comments about the practical aspects of this task are in order. For example, there is a basic assumption that the trades suggested by the system can be executed. If we have a 2000 stock universe and hope to develop a market-neutral portfolio, we would have to be long in some stocks and short in others. The above analysis is based upon the top and bottom 10 percent of the universe of stocks which means that for any given month the portfolio would be long for approximately 200 stocks and short for approximately 200 others. Furthermore, there is an implicit assumption that the position size for all the stocks is approximately equal (on a dollar basis). A practical problem would arise if one attempts to short all the 200 stocks in the bottom group. For lower capitalization stocks in this group, shorting presents an execution problem. One possible solution is to alter the amount bet on each stock as a percentage of its total capitalization. An alternative is to limit the universe to only large capitalization stocks. However, if either of these approaches is followed, the analysis developed above isn't strictly applicable. This is one example of the types of problems encountered in turning a modeling strategy into a trading system. The practical realities of trading ultimately must be faced when the time comes to turn a good modeling strategy hopefully into a successful trading system.

NOTES

1. P. J. Kaufman, *Trading Systems and Methods,* 3rd ed. (New York: John Wiley & Sons, 1998).

2. J. Schwager, *Schwager on Futures: Technical Analysis* (New York: John Wiley & Sons, 1996).

3. L. Williams, *Long-Term Secrets to Short-Term Trading* (New York: John Wiley & Sons, 1999).

4. R. Pardo, *Design, Testing and Optimization of Trading Systems* (New York: John Wiley & Sons; 1992).

5. J. P. O'Shaughnessy, *What Works on Wall Street* (New York: McGraw-Hill, 1997).

6. M. A. Ruggiero, Jr., *Cybernetic Trading Strategies* (New York: John Wiley & Sons, 1997).

7. M. Jurik, *Computerized Trading* (New York: New York Institute of Finance, 1999).

8. T. S. Chandee and S. Kroll, *The New Technical Trader* (New York: John Wiley & Sons, 1994).

9. T. S. Chandee, *Beyond Technical Analysis* (New York: John Wiley & Sons, 1997).

10. D. Aronson, "Pattern Recognition Signal Filtering," *Journal of the Market Technicians' Association* (Spring 1991).

APPENDIX A

LINEAR LEAST SQUARES WITH DATA WEIGHTING

The method of least squares is used to estimate parameters in a function of the form:

$$y = f(x_1, x_2..., x_m; a_1, a_2..., a_p) \tag{A.1}$$

In this equation y is a function of m independent variables x_1 to x_m and p parameters a_1 to a_p. In the more general case, y can be a vector and f will then be a set of functions. Furthermore, the function (or functions) f might be nonlinear. The application of the method of least squares in this book is limited to linear functions:

$$f(x_1, x_2.., x_m; a_1, a_2.., a_p) = \sum_{j=1}^{j=p} a_j \, g_j \, (x_1, x_2.., x_m) \tag{A.2}$$

Using vector notation:

$$f(X, A) = \sum_{j=1}^{j=p} a_j \, g_j \, (X) \tag{A.3}$$

where X is the vector of the independent variables and A is the vector of unknown parameters. The purpose of method of least squares is to provide an estimate of the p unknown parameters that minimize the least squares criterion S:

$$S = \sum_{i=1}^{n} w_i (Y_i - f(X_i, A))^2 \qquad \text{(A.4)}$$

In this equation Y_i is the actual value of Y for the ith data point and X_i is the vector of independent variables for the ith data point. The term w_i is the weight associated with the ith data point.

Substituting Eq. (A.3) into (A.4) and simplifying the notation, we obtain the following equation:

$$S = \sum_{i=1}^{n} w_i \left(Y_i - \sum_{j=1}^{j=p} a_j g_j \right)^2 \qquad \text{(A.5)}$$

To determine the elements a_k of the parameter vector A, the p derivatives of S with respect to each of the a_k's is set to zero. Each derivative results in a linear algebraic equation, and the p equations are then solved for the p values of a_k:

$$\frac{\partial S}{\partial a_k} = -2 \sum_{i=1}^{n} w_i g_k \left(Y_i - \sum_{j=1}^{j=p} a_j g_j \right) = 0 \qquad \text{(A.6)}$$

Rearranging this equation, we obtain p equations of the form:

$$\sum_{j=1}^{j=p} a_j \sum_{i=1}^{n} w_i g_k g_j = \sum_{i=1}^{n} w_i g_k Y_i , \quad k = 1 \text{ to } p \qquad \text{(A.7)}$$

This set of equations can be expressed in matrix form as follows:

$$CA = V \qquad \text{(A.8)}$$

In this equation C is a p by p matrix and A and V are vectors of length p. The C matrix is symmetrical, and the term C_{jk} is defined as follows:

$$C_{jk} = \sum_{i=1}^{i=n} w_i g_j g_k \qquad \text{(A.9)}$$

The term V_k is defined as follows:

$$V_k = \sum_{i=1}^{i=n} w_i g_k Y_i \qquad (A.10)$$

Solution of Eq. (A.8) can be based on any number of well-known methods for solving a set of linear equations. If, however, the equations are solved by first inverting the C matrix, then the terms of the inverse matrix C^{-1} can be used to estimate errors associated with Eq. (A.3). It can be shown that the estimated variance of the fitted function f at any point in the X space is:[1,2]

$$\sigma_f^2 = \frac{S}{n-p} \sum_{j=1}^{p} \sum_{k=1}^{p} g_j g_k C_{jk}^{-1} \qquad (A.11)$$

Only the values of the functions g_j and g_k vary from point to point. So once the terms of the C matrix and V vector have been determined, the terms of the A vector can then be determined by solving the matrix Eq. (A.8). Estimates for Y at any point can be made using Eq. (A.3). (The function f is the least squares estimator for Y.) An estimate of the standard deviation associated with f may be made by taking the square root of Eq. (A.11).

NOTES

1. J. Wolberg, *Prediction Analysis* (New York: Van Nostrand, 1967).

2. P. Gans, *Data Fitting in the Chemical Sciences* (New York: John Wiley & Sons, 1992).

APPENDIX B

A TEST FOR SIGNIFICANCE OF VARIANCE REDUCTION

In Section 6.2, expressions for a test of significance for VR (Variance Reduction) are included (Eqs. 6.1 to 6.3). In this appendix, these equations are derived.[1] This derivation is applicable to the *Order 0* Algorithm in which the kernel smoothing parameter as defined by Eq. (3.1) is 0 (i.e., all points are equally weighted). For this case the average values of all learning points in each cell are used as predictors for the dependent variable y.

In Eq. (1.12) VR was defined as a percentage. In this appendix it is more convenient to define it as a fraction, as follows:

$$VR = \frac{V_{tt} - V_{tl}}{V_{tt}} \tag{B.1}$$

The V's are sums. V_{tt} is the sum of the squares of the differences between the test values of y and the average test value μ_t. V_{tl} is the sum of the squares of the differences between the test values and the average value of the learning points in the N_c cells:

$$V_{tt} = \sum_{i=1}^{N_t} (yt_i - \mu_t)^2 \tag{B.2}$$

$$V_{tl} = \sum_{k=1}^{N_c} \sum_{j=1}^{N_{tk}} (yt_{kj} - \mu_{lk})^2 \tag{B.3}$$

In these equations yt_i is the value of y for the ith test point, and yt_{kj} is the value for the jth test point in cell k. The mean value of the learning points in cell k is denoted as μ_{lk}. The number of test points in cell k is N_{tk}, and the number of learning points is N_{lk}, which is essentially a constant value (due to the manner in which the learning points are partitioned into cells). To simplify the analysis, we assume that the number of test points is equal to the number of learning points and, furthermore, that the number in each cell is equal to N_d. This is not a necessary assumption, but it does make the analysis simpler.

We assume that the data has no structure (i.e., the values of μ_{lk} are not useful predictors for the values of y for the test points). Using this assumption and the assumption of a constant value of N_d test and learning points in each cell (i.e., $N_d = N_v/N_c$ where $N_v = N_t = N_l$), we can compute the expected distribution for VR. To do this, the test and learning points are partitioned randomly into the N_c cells. For each cell we have:

$$\{yl_{kj} : j + 1, \dots N_d\}, \{yt_{kj} : j = 1, \dots N_d\} \qquad (B.4)$$

Under these assumption, we can compute a distribution for the mean values of the learning point average. It can be shown that under mild conditions on the y sequences, the distributions of μ_{lk} and μ_{tk} are approximately multivariate normal with mean vectors μ_l $(1,\dots,1)$ and μ_t $(1,\dots,1)$. The covariance matrices are $(\sigma_v)^2(I - J/N_c)/N_d$, where the subscript v is either l or t (for the learning or test distribution).[2] The symbol I represents the identity matrix, J represents a matrix full of 1's, and both matrices are N_c by N_c. The symbols σ_l and σ_t are the standard deviations of the distributions.

Using an analysis very similar to the analysis of variance, we can modify Eq. (B.3):

$$V_{tl} = \sum_{k}^{N_c} \sum_{j}^{N_d} (yt_{kj} - \mu_t + \mu_t - \mu_{tk} + \mu_{tk} - \mu_{lk})^2 \qquad (B.5)$$

Expanding the term inside the summation leads to the following expression:

$$V_{tl} = V_{tt} - \sum_{k=1}^{N_c} N_d (\mu_{tk} - \mu_t)^2 + \sum_{k=1}^{N_c} N_d (\mu_{tk} - \mu_{lk})^2 \qquad (B.6)$$

From this equation and after some manipulation, the following expression is obtained:

$$V_{tt} - V_{tl} = -N_d N_c (\mu_t - \mu_l)^2 - N_d u' H u \tag{B.7}$$

In this equation u is a $2N_c$ vector $\{\mu_{t1}-\mu_t, \mu_{t2}-\mu_t,\dots, \mu_{l1}-\mu_l, \mu_{l2}-\mu_l,\dots\}$. H is a $2N_c$ by $2N_c$ matrix:

$$H = \begin{bmatrix} I & -I \\ -I & 0 \end{bmatrix} \tag{B.8}$$

We are now in a position to use the covariance matrixes of the cell mean values. Defining A as the N_c by N_c matrix $(I\text{-}J/N_c)$, we find that the exact covariance matrix of $sqrt(N_d)u$ is:

$$\Sigma = \begin{bmatrix} \sigma_l^2 A & 0 \\ 0 & \sigma_t^2 A \end{bmatrix} \tag{B.9}$$

From this equation we can estimate the expected value of $V_{tt} - V_{tl}$:

$$E(V_{tt} - V_{tl}) = -N_d N_c (\mu_t - \mu_l)^2 - tr(H\Sigma) \tag{B.10}$$

$$E(V_{tt} - V_{tl}) = -N_d N_c (\mu_t - \mu_l)^2 - \sigma_l^2 (N_c - 1) \tag{B.11}$$

To obtain $E(VR)$, the expected value of VR, we can use the following definition of variance:

$$\sigma_t^2 = \sum_{t=1}^{N_t} (yt_i - \mu_t)^2 /(N_t - 1) = V_{tt} /(N_t - 1) \tag{B.12}$$

From (B.11), (B.12), and (B.1) we get the following expression for $E(VR)$:

$$E(VR) = -\frac{N_t}{N_t - 1}(\mu_t - \mu_l)^2 / \sigma_t^2 - \frac{(N_c - 1)\sigma_l^2}{(N_t - 1)\sigma_t^2} \qquad \text{(B.13)}$$

For the cases where the test and learning points are from the same population, and $N_t = N_l$ is large, we can assume $\mu_t \approx \mu_l$ and $\sigma_t \approx \sigma_l$. Equation (B.13) reduces to the simple form:

$$E(VR) = -\frac{(N_c - 1)}{(N_t - 1)} \qquad \text{(B.14)}$$

The distribution of VR depends on that of the quadratic form $N_d u' H u$. Asymptotically, as $N_t = N_l$ approaches ∞, N_c fixed, $sqrt(N_d)u$ is multivariate normal with mean 0 and covariance matrix Σ. An eigenanalysis of $H\Sigma$ shows that it has eigenvalues given by $(\sigma_t)^2 \lambda_1$ and $(\sigma_t)^2 \lambda_2$:

$$\lambda_{1,2} = \left(1 \pm \sqrt{(\sigma_l / \sigma_t) + 4}\right) / 2 \qquad \text{(B.15)}$$

each of multiplicity (N_c-1) and a double eigenvalue of 0. Writing Q_1 and Q_2 for two independent chi-squared variables, each of (N_c-1) degrees of freedom, we conclude that asymptotically $(N_t$ large and N_c fixed) VR is distributed as follows:

$$VR = -\frac{N_t}{N_t - 1}(\mu_t - \mu_l)^2 / \sigma_t^2 - \frac{1}{(N_t - 1)}(\lambda_1 Q_1 + \lambda_2 Q_2) \qquad \text{(B.16)}$$

From this equation we may deduce that:

$$var(VR) \approx 2(N_c - 1)(\lambda_1^2 + \lambda_2^2) / (N_t - 1)^2 \qquad \text{(B.17)}$$

$$var(VR) \approx (5 + (\sigma_l/\sigma_t)^4)(N_c - 1)/(N_t - 1)^2 \qquad \text{(B.18)}$$

Finally, for $\sigma_t \approx \sigma_l$:

$$var(VR) \approx 6(N_c - 1)/(N_t - 1)^2 \qquad \text{(B.19)}$$

To compute $C(VR)$, the critical values of VR based on (B.16), there are methods for approximating quantiles of linear combinations of chi-squared variables. However, we have found that a normal approximation is adequate and even on the conservative side. Taking the mean and variance of VR from Eqs. (B.13) and (B.18), or more simply from (B.14) and (B.19) we use the following simple expression:

$$C(VR) = \mu(VR) + \eta\sigma(VR) \qquad (B.20)$$

The multiplication factor η in this equation is the number of standard deviations used to set the critical value VR and can be taken from a standard normal table. For example, if we want to set the value at a point where the probability of exceeding $C(VR)$ is only 1 percent then the value of η will be set at 2.33.

As an example, assume $N_t = N_l = 5000$ and the treeheight used to partition the learning data is 7. The value of N_c is $2^7 = 128$ and therefore from (B.14) $\mu(VR) = -127/4999 = -0.0254$. From (B.18) we get $\sigma(VR) = sqrt(6*127/4999^2) = 0.0055$. From (B.20) using $\eta = 2.33$ we get $C(VR) = -0.0225 + 2.33*0.0055 = -0.0126$. Note that a value of $VR = 0$ is actually significant! Note also that significance does not necessarily mean useful. As another example, assume a treeheight of 2 (i.e., $N_c = 4$) and $N_t = N_l = 200$. $\mu(VR) = -3/199 = -0.0050$, $\sigma(VR) = sqrt(6 * 3/199^2) = 0.0213$, $C(VR) = -0.0050 + 2.33*0.0213 = 0.0446$. In other words, VR must exceed 4.46 percent to pass this significance test.

NOTES

1. P. D. Feigin and J. R. Wolberg, "A Test of Significance for the Cells Method of Screening in Nonparametric Regression" (Unpublished, 1999).

2. A. Wald and J. Wolfowitz, "Statistical Tests Based on Permutations of the Observations" *Ann. Math. Statistics* 15 (1984): 358–372.

APPENDIX C

COMPARING KR AND PARAMETRIC REGRESSION

The most well-known parametric method for data modeling is the method of *least squares*. In Appendix A the method is developed for cases in which the relationship between the dependent variable y and the m independent variables x_1 to x_m is linear with respect to the p parameters a_1 to a_p of the model. Linearity is not a limitation of this method. More generally the relationship may be nonlinear, and the dependent variable y may be a vector.[1-3] Easily usable software is available for general nonlinear least squares. I have been involved in the development of a package called REGRESS that is now available as a product of Insightware Ltd. (*www.insightware.com*) Other nonlinear regression packages can be located through the Internet site *www.kdnuggets.com*.

For many scientific and engineering problems, a parametric approach to modeling is essential. Often, the purpose of the model is to extract parameters by analyzing the data. In Section 1.1 a simple experiment for analyzing the count rate of a radioisotope as a function of time was discussed. Putting Eq. (1.3) in a slightly different form, we have a nonlinear equation relating count rate y and time t:

$$y = a_1 e^{-(a_2 * t)} + a_3 \qquad \text{(C.1)}$$

Typically, the purpose of such experiments is to determine the decay constant a_2. Usually, the analyst is interested in an estimate of the uncertainly associated with the experiment. For this example, the estimated value of σ_{a2} (the standard deviation of a_2) would be of particular interest.

The problem with financial market modeling is that a mathematical equation relating y (the variable to be modeled) and the x's (the candidate predictors) is rarely available. However, when such an equation is obtainable, the amount of information potentially available to the analyst is impressive. To compare KR and the method of *least squares*, the following procedure was followed:

1. Create an artificial data set based on a known nonlinear relationship.
2. Corrupt the data with random noise.
3. Analyze the data using REGRESS (a nonlinear least squares software package).
4. Analyze the data using FKR (*www.insightware.com*).

A TIMES program (Figure C.1) was written to generate data based on the following equation:

$$f(x) = a_3 * \exp(-(((x - a_1)/a_2)^2)) - a_6 * \exp(-(((x - a_4)/a_5)^2)) \quad (C.2)$$

```
proc main() {
  n = 15000;
  a = reshape([n,6],0);  // creates an n by 6 zero filled matrix
  x = grnum(n);          // x is vector of Gaussian random numbers
  a[:1] = x;             // fill col 1 of matrix with x
  y = exp(-((x + 0.5)^2)) - exp(-((x - 0.5)^2));
  a[:2]=y;   // Col 2: The pure signal, a 1D non-linear function

  noise=rnum(n);
  sran=sigma(noise);
  sy=sigma(y);
  a[:3] = y + (sy/sran)*sqrt(0.50/0.50)*noise;      // maxvr=50
  a[:4] = y + (sy/sran)*sqrt(0.75/0.25)*noise;      // maxvr=25
  a[:5] = y + (sy/sran)*sqrt(0.90/0.10)*noise;      // maxvr=10
  a[:6] = y + (sy/sran)*sqrt(0.95/0.05)*noise;      // maxvr= 5

  datecol = 0;   // No need for a date column
  boutput("appendixc.pri", datecol, a);  // binary output file
  printf("appendixc.pri created");
}
main();
```

Figure C.1 TIMES program for generating test data set.

The values selected for the six parameters were $a_2 = a_3 = a_5 =$ $a_6 = 1$, $a_1 = -0.5$, and $a_4 = 0.5$. This function is shown in Figure C.2. The function is antisymmetric about $x = 0$ with a maximum of 0.73 at about -0.8 and a minimum of -0.73 at about 0.8. The y value used for modeling was created by adding random noise to $f(x)$. Values of x were selected from a Gaussian normal distribution (mean 0 and σ of 1). The resulting values of y chosen for analysis were from column 5 of the data file. This column was generated such that the true signal was corrupted with a 90 percent random noise component.

Results from the REGRESS analysis are shown in Figure C.3. The initial guess used in the REGRESS analysis for the position of the maxima (i.e., a_1) was set at -1, and the initial guess for the minima (i.e., a_4) set at 1. After five iterations, the convergence criterion was satisfied, and the resulting values of a_1 and a_4 are seen to be -0.768 and 0.770. These values are significantly different from the parameters used to generate the pure signal (i.e., -0.5 and 0.5), but this is understandable because the column analyzed (column 5) was generated using a 90 percent noise component. Notice also that the other four parameters are also different from the actual values used to generate the pure signal.

Results from the FKR analysis are shown in Figure C.4. For this particular run the *treeheight* was 4 and the *order* was 1,

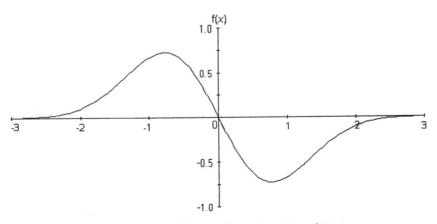

Figure C.2 $f(x)$: Nonlinear function of X.

```
PARAMETERS USED IN REGRESS ANALYSIS: Wed Jan 20 13:07:21 1999
  INPUT PARMS FILE: appendixc.par
  INPUT DATA  FILE: appendixc.pri
  Total number recs read from data file   : 15000
  STARTREC - First record used            :     1
  N - Number of recs used to build model  : 10000
  STARTEVAL - First record in eval set    : 10001
  NEVL - Number of recs in eval data set  :  5000
  NO_DATA - Code for dependent variable    -999.0
  NCOL - Number of data columns           :     6
  YCOL1 - Column for dependent variable 1 :     5
  SYTYPE - Sigma type for Y               :     1
     TYPE 1:  SIGMA Y = 1
  M - Number of independent variables     :     1
  XCOL - Column for X                     :     1
  SXTYPE - Sigma type for X               :     0
     TYPE 0:  SIGMA X = 0

Analysis for Set 1
  Function F1: A3*EXP(-(((X-A1)/A2)^2))-A6*EXP(-(((X-A4)/A5)^2))
  EPS - Convergence criterion             : 0.00100
  CAF - Convergence acceleration factor :   1.000

ITER      A1       A2       A3       A4       A5       A6  S/(N.D.F.)
   0  -1.0000   1.0000   1.0000   1.0000   1.0000   1.0000   2.49061
   1  -0.7610   0.9738   0.6963   0.7558   0.9610   0.7204   2.43934
   2  -0.8340   0.7693   0.6866   0.8353   0.7661   0.7094   2.43708
   3  -0.7845   0.8442   0.7172   0.7837   0.8337   0.7409   2.43709
   4  -0.7688   0.8447   0.7289   0.7705   0.8344   0.7536   2.43703
   5  -0.7680   0.8466   0.7292   0.7695   0.8354   0.7544   2.43703

 K      AO(K)      AMIN(K)      AMAX(K)       A(K)       SIGA(K)
 1   -1.00000    Not Spec    -0.30000    -0.76821     0.10108
 2    1.00000     0.20000     Not Spec    0.84620     0.13693
 3    1.00000    Not Spec    Not Spec     0.72919     0.04560
 4    1.00000     0.30000    Not Spec     0.76972     0.09792
 5    1.00000     0.20000    Not Spec     0.83512     0.12720
 6    1.00000    Not Spec    Not Spec     0.75429     0.04806

Variance Reduction:           9.43
S/(N - P)          :          2.43703
RMS (Y - Ycalc)    :          1.56063

Evaluation of Model for Set 1:
  Number of points in evaluation data set:   5000
  Variance Reduction                        10.56
  RMS (Y - Ycalc)                            1.54578
  Fraction Y_eval positive                : 0.505
  Fraction Y_calc positive                : 0.494
  Fraction Same Sign                      : 0.604
     This value of FSS is significant at level  0.05

Data Set  Variable  Minimum   Maximum   Average   Std_dev
Modelling    X      -3.7380    3.7750    0.0067    0.9977
Modelling    Y      -3.4027    3.4257   -0.0014    1.6400
Evaluate     X      -3.4376    2.9713    0.0105    0.9786
Evaluate     Y      -3.4270    3.4162    0.0035    1.6346
```

Figure C.3 Output from REGRESS analysis of *appendixc.pri* data file.

```
Output report:   FKR Analysis Wed Jan 20 16:28:03 1999
                 Version 3.11  Jan 17, 1999
Parameters used in analysis    PARAMETER FILE: appendixc.fpa
                               INPUT DATA FILE: appendixc.pri


DIM     - number of dimensions                   :       1
NCP     - number of candidate predictors         :       1
NCOL    - number of columns of data              :       6
NREC    - total number of records                :   15000
LTYPE   - Learning set type (G, M or S)          :       S
          G is growing, M is moving and S is static
NUMCELL- Number of cells searched for NUMNN       :       1
NUMNN   - Irrelevant because FAST mode specified
STARTVAR - starting variable                      :       1


Kernel Regression parameters:
MOD_CRITERION- 1=VR, 2=CC, 3=CC_O, 4=Separation :       1
FIT_ORDER - 0 is average,  1 & 2 are surfaces    :       1
    FAST option used: Ycalc from cell hyperplane
NUMK    - Number of smoothing parameters          :       1
K(1)    - Smoothing parameter 1                   :    1.00
YCOL    - Column of dependent variable            :       5


Tree parameters:
BUCKETSIZE - design number per cell (computed)   :     625
TREEHEIGHT - tree parm                            :       4
Computed number of cells                          :      31
Computed number of leaf cells                     :      16
Computed avg bucket size                          :   625.0
Computed 2 Sigma limit for F Stat                 :    1.73


Data set parameters:
NLRN    - number of learning records              :   10000
NTST    - number of test records                  :    5000
NEVL    - number of evaluation records            :       0
GAP     - gap records between data sets           :       0
STARTLRN - starting learning record               :       1
STARTTST - starting test record                   :   10001

Ordered results for 1 dimensional models:
    Total number of combinations:         1
    Number tested          :              1
    survivenum( 1)         :              1
    Number of survivors    :              1
Var Red:   10.403   FracSS: 0.604    F: 69.09   X:   1
```

Figure C.4 Output from FKR analysis of *appendixc.pri* data file.

however, the results were insensitive to both of these parameters. What is interesting to note is that the value of VR (Variance Reduction) was essentially the same for both the REGRESS and FKR runs (10.56 for REGRESS for the 5000 evaluation data cases and 10.40 for FKR for the 5000 test cases). For this particular data set, both methods were extremely efficient and achieved slightly more than the maximum variance reduction that one could expect from data that was created by adding a 90 percent random noise component. Note that in REGRESS a value of 9.43 is also listed for VR, but this value is for the 10,000 data points used to determine the six parameters of the model. The resulting models using these two totally different methods yield results that are quite similar.

Comparing the two results, it is clear that the REGRESS analysis yields more information than the FKR analysis. The analyst not only has a model that can be used to predict y for any value of x, but also has an analytical form of the model with values and error estimates for the six parameters included in the model (i.e., Eq. C.1). For many engineering and scientific applications, models of this type are the main object of the analysis. However, for financial market modeling, such analytical models, rarely, if ever, exist. For this example, the model is one dimensional; however, for higher dimensional models, more parameters are required. For example, a similar there-dimensional artificial data set was created using the TIMES program shown in Figure 5.1. For that particular data set, 18 parameters would be required to specify an analytical model similar to Eq. (C.1). As the number of parameters increases, convergence for nonlinear least squares becomes more of a problem.

NOTES

1. Y. Bard, *Nonlinear Parameter Estimation* (San Diego, Cal: Academic Press, 1974).

2. P. Gans, *Data Fitting in the Chemical Science* (New York: John Wiley & Sons, 1992).

3. J. Wolberg, *Prediction Analysis* (New York: Van Nostrand Reinhold, 1967).

APPENDIX D

COMPARING KR AND NEURAL NETWORKS

Currently, the most popular nonparametric method for data modeling is probably the method of neural networks. The literature on NN (neural networks) is vast and growing. For a detailed description of the many different approaches to this subject, a book by Haykin is often the primary reference for many of the authors writing about NN.[1] A more recent second edition of this book was published in 1998. Other books of general interest include books by Patterson and Smith.[2-3] Several books are available on the usage of NN in financial applications.[4-7] A recent survey of modeling methodologies in the *PC AI Journal* places the major emphasis on neural networks.[8]

Modeling financial markets typically involves a large number of candidate predictors. Although some people feel that it is possible to model markets using a small number of clever indicators, most people agree that such "magic bullets" are hard to find. As a result, the typical approach is to propose a fairly large number of candidate predictors and then search for a subspace with adequate predictive power. Theoretically, neural networks can be designed to accommodate a large number of candidate predictors. If the process is successful, one would expect that weights associated with irrelevant candidate predictors will approach zero and only the weights of the candidate predictors with real predictive power will be significant. The problem with this approach is the large amount of calculational time required to achieve a good model and the increasing probability of getting stuck on a local minimum as the number of inputs increases.

The underlying mathematical problem that NN attempts to solve is nonlinear with many unknowns. This is not an easy task. One difficulty encountered when trying to solve a set of nonlinear equations is the problem of local minima. The design of NN algorithms must include some method for overcoming this problem. One well-known approach to the problem is known as the *Boltzmann learning rule*.[9-10] This method is named after the famous thermodynamicist, L. Boltzmann, and is similar to the approach he used to determine an energy state of a thermodynamic system. In the neural net implementations, random variations in the states of some neurons in the network are introduced, and through this device sometimes the system is jarred out of a local minimum. A successful method for achieving rapid convergence is the *Levenberg–Marquardt Algorithm*.[11-13] This algorithm is based on early work by K. Levenberg[14] and D. W. Marquardt.[15] Rapid implementation of this algorithm is included in the MATLAB™ *(www.math-works.com)* Neural Network Toolbox.[16]

Typically, the network is "trained" using a training (or learning) data set. A network structure is proposed, and based on this structure, a number of free parameters are used to relate the outputs to the network inputs. For example, Figure D.1 shows a network with R inputs, and three layers of adaptive neurons with $S1$, $S2$, and $S3$ neurons, respectively. This figure comes from the MATLAB Neural Network Toolbox Users Guide.[16] The number of weights for this network is R per each neuron in the first layer, $S1$ per neuron in the second layer, and $S2$ per neuron in the third layer. In addition, the summing operation performed at the input of each neuron is biased by an adaptive bias parameter. Thus the total number of free parameters (weights and biases) is $(R + 1)S_1 + (S_1 + 1)S_2 + (S_2 + 1)S_3$.

The f blocks shown in Figure D.1 are nonlinear transfer functions, and most neural network software packages include a variety of different functions that may be selected. The biases associated with the summing operations are also shown in Figure D.1. Starting from a random selection for the free parameters, an iterative search is initiated to determine a set of values that answer a specified "stopping" criterion. A typical stopping criterion is based on the sum of the squared errors. When this value becomes less than a specified amount or when the change in this value is less than a specified fraction, the process is ter-

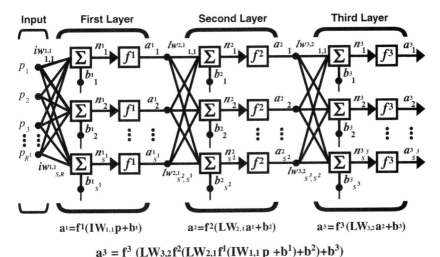

$$a^1 = f^1(IW_{1,1}p + b^1)$$

$$a^2 = f^2(LW_{2,1}a^1 + b^2)$$

$$a^3 = f^3(LW_{3,2}a^2 + b^3)$$

$$a^3 = f^3 \left(LW_{3,2}f^2(LW_{2,1}f^1(IW_{1,1}\,p + b^1) + b^2) + b^3\right)$$

Figure D.1-A   Neural network with R inputs, 3 layers and 3 neurons per layer.[16]

minated. To avoid the possibility of an infinite loop, a maximum number of iterations (or *epochs*) is usually specified.

Even though the NN approach to modeling is extremely popular, for some classes of problems this approach has serious drawbacks. In particular, problems with a large number of candidate predictors are extremely difficult to solve using a standard NN approach. The standard approach is to submit all the candidate predictors to the network and then allow the process to determine the relative importance of each candidate predictor by adjusting the weights. The term "large" is of course relative and depends on the available environment. However, some authors have indicated that the calculational complexity of an NN analysis is exponentially dependent on the number of inputs.[17–18] Others feel that the complexity grows in a polynomial manner.[19] Results from our timing experiments are discussed below. If the number of candidate predictors is indeed large, then a search strategy similar to the one described in Sections 3.6 and 4.7 is essential.

To compare KR and NN, a typical test problem was created. The problem is based on a data set similar to the data set used in Chapter 5. A pure signal was created using the following equations:

$$yc[i] = \exp(-((x[i] + 0.5)^\wedge 2))$$
$$- \exp(-((x[i] - 0.5)^\wedge 2)) \quad i = 1,2,3 \qquad \text{(D.1)}$$

$$y = yc[1]^*yc[2]^*yc[3] \qquad \text{(D.2)}$$

Each of the three x vectors included 15000 Gaussian random numbers (i.e., mean 0 and $\sigma = 1$). For this data set, the average value of y is close to zero, and there are four peaks and four valleys in the three-dimensional space created by the three x variables. The resulting y vector was then corrupted with varying levels of noise using Eq. (5.1). The results in Tables D.1 and D.2 are for the combined signal plus noise vector with a noise component of 90 percent. The MATLAB Neural Network Toolbox (*www.mathworks.com*) was used to determine the NN results. In the NN Toolbox Users Guide, the authors recommend using the routine TRAINLM to train the network. Indeed, their comparative statistics show that this routine (based on the Levenberg–Marquardt algorithm) outperforms the alternative routines by a substantial amount.[16] The results presented below are based upon a modified form of TRAINLM.[20] The stopping criterion used in this routine is when the fractional change in the sum of the squared errors becomes less than a user-specified amount. We used a value of 0.001. The number of inputs R to the network was 3 (i.e., the three x columns). The network included a hidden layer of $S1=$ *num_neurons* with all three inputs used for all the hidden neurons. Referring to Figure D.1, the network also included a single-output neuron at the output layer (i.e., $S2 = 1$ and $S3 = 0$). The total number of free parameters is therefore $(R + 1)S_1 + (S_1 + 1) = (R + 2)S_1 + 1 = 5^*num\_neurons + 1$.

The results in Table D.1 were obtained by using the first 7500 learning points to train the network, and then the resulting network was tested using the remaining 7500 test points. (In other words, *nlrn* = *ntst* = 7500.) The number of epochs, time (in seconds), and resulting variance reduction (VR) obtained using the 7500 test points were recorded for each of eight trials for each of the four values of *num_neurons*. For a given value of *num_neurons,* the results varied from trial to trial due to the random selection of the initial free parameters. For example, for the case of 10 neurons, the number of epochs required to achieve the stopping criterion varied from 10 to 14 with an average value of 11.87. Since the time required per epoch is almost constant, the variation in time is due almost entirely to the number of epochs required to achieve convergence. The variation in VR is due to the random selection of initial free parameters. The environment used for these experiments was a Pentium PII-400 with 128 Meg of RAM running under NT.

An analysis of the results in Table D.1 shows that for this particular data set the best results are obtained for *num_neurons* = 20. The range of average values of VR is from 6.65 to 7.74 percent. Since the data is only 10 percent signal and 90 percent noise, we see that the NN results capture approximately 70 percent of the available signal. The average *Time_per_Epoch* for the five different values of *num_neurons* increases at a rate that is greater than O(*num_neurons*) but less than O(*num_neurons*^{2}). An attempt was made to measure the time required per test point by reducing *ntst* to a small number; however, there was no noticeable difference in the time per epoch. In other words, using the terminology of Eq. (4.5), we find that the value of T_{avg} (the average time to analyze a space) is almost entirely equal to T_{prep} (i.e., preparation time) and T_{run} (the running time to test the model) is essentially zero. For this example, once the network has been trained, the time to test it is negligible.

The data was then analyzed using the FKR program (*www.insightware.com*) and the same hardware. Results are included in Table D.2. There is no random component to a KR analysis, so the experiments were only performed once for each combination of parameters. All the results in this table were

TABLE D.1 Results for a Neural Network Experiment Using *nlrn* = 7500 and *ntst* = 7500

Num_Neurons	Avg_Epochs	Avg_Time (secs)	Avg_Time/ Epoch	Avg_VR	Sigma_VR
10	11.87	22.74	1.916	6.65	0.30
15	12.80	40.07	3.130	6.99	0.90
20	11.30	53.57	4.741	7.74	0.25
30	10.38	87.37	8.417	7.46	0.37
40	9.25	135.71	13.590	7.33	0.61

obtained using a treeheight $h = 7$ (*i.e.*, $2^7 = 128$ cells). The parameters specifying the nearest neighbor search are *numnn* (the number of nearest neighbors) and *numcells* (the number of cells in which the searches are to be performed). Regardless of the value of *numcells,* for each test point the search is limited to the test cell and all adjacent cells. By setting *numcells* to a large value, we can be assured that all adjacent cells are included in each search. The table includes results for the two extreme types of analysis: (i) the *fast option* in which the nearest neighbors are just the points in the test cells, and (ii) the most complete nearest neighbor search, including all adjoining cells to every test cell. We see that, as expected, the recorded times for option *ii* are much greater than for option *i*. Regardless of the *order* of the algorithm used to make predictions for the values of *Y* for the test points, the total time for a fast analysis is approximately 0.1 second. This is orders of magnitude faster than the NN results in Table D.1, but the resulting values of VR are significantly less. In fact, for *order* = 2, the value of *VR* is negative. To obtain comparable *VR* using *order* = 2, a full nearest neighbor search is required.

TABLE D.2 Results for a Kernel Regression Experiment Using *nlrn* = 7500 and *ntst* = 7500

Order	numnn	numcells	VR	Time (secs)
0	all in cell	1	5.16	0.08
0	25	all adjacent cells	6.15	2.60
0	50	all adjacent cells	7.13	2.85
0	100	all adjacent cells	7.45	3.90
0	150	all adjacent cells	7.33	5.20
1	all in cell	1	3.24	0.09
1	50	all adjacent cells	5.51	3.10
1	100	all adjacent cells	7.51	4.30
1	150	all adjacent cells	7.71	5.80
1	200	all adjacent cells	7.53	7.50
2	all in cell	1	−12.84	0.12
2	150	all adjacent cells	4.59	8.20
2	300	all adjacent cells	6.94	15.00
2	500	all adjacent cells	7.63	23.00
2	600	all adjacent cells	7.45	26.00

The value of T_{prep} (the preparation time to create the models) was measured by setting $ntst$ to a very small value. The results for KR are totally different than for NN: the value of T_{prep} was less than 0.1 second for this problem. In other words, if a full nearest neighbor search is used in a KR analysis for problems of this size, the time to analyze a space is almost entirely T_{run} and therefore is essentially linearly dependent on the number of test points, that is, O($ntst$). On the other hand, an NN analysis is essentially independent of $ntst$. Some other interesting conclusions can be made from the results shown in Table D.2. The optimum value of $numnn$ increases with increasing $order$ for the full nearest neighbor search: the best results for $order$ = 0, 1 and 2 were obtained for $numnn$ = 100, 150 and 500. Since there were only about 59 (i.e., 7500/128) learning points per cell, the fast option for $order$ = 2 suffers from a lack of sufficient points to make accurate predictions.

From the results shown above for this particular problem, it is not clear which approach (KR or NN) would yield the best model from the point of view of VR. The best results for both techniques captured over 70 percent of the available 10 percent signal. However, even if one decides that NN is preferable, it is obvious that if the number of candidate predictors is large, then the *fast* KR approach using the *Order* 0 Algorithm should be used for parameter screening and selection. For problems of the size of the test problem but with many candidate predictors, the ability to analyze a space in about 0.1 second allows the analyst to scan thousands of spaces in a reasonable amount of time. Once the best candidate predictors (or combinations of candidate predictors) have been located, then one can proceed to an NN analysis or to a more accurate nearest neighbor KR analysis.

An additional experiment was designed to measure the time dependence of the neural network software. The number of candidate predictors (i.e., ncp) was varied from 1 to 10 using the data from the same three input model as used to determine the results in Tables D.1 and D.2 (i.e., Eqs. D.1 and D.2). The parameters that were varied besides ncp were *num_neurons* (from 5 to 40), *nlrn* (from 100 to 7500), and the number of *epochs* (from 5 to 30). Using the REGRESS program (*www.insightware.com*) with constant fractional error weighting for the *Time* values, the following equation for *Time* (in seconds) was determined from 76 combinations of parameters:

$$Time = 0.61 + (nlrn/1000) * (4.01 * 10^{-4} * ncp^{1.16} *$$
$$num\_neurons^{1.93} * epochs + 2.64) \qquad (D.3)$$

For example, for *nlrn* = 7500, *ncp* = 3, *num_neurons* = 10 and *epochs* = 20, this equation yields a value of *Time* = 38.7 seconds. This equation is valid only for the particular environment in which the experiment was performed (i.e , a Pentium II-400 with 128 Meg of RAM running under NT). The root mean square fractional error between the calculated and actual values of *Time* was about 13 percent. The interesting point to note is that the exponential behavior of the *ncp* term is very weak. In fact, it is nearly linear.

Besides the timing results, this experiment showed that even when *ncp* exceeds 3 (i.e., some extraneous input variables are included), the value of VR (variance reduction) does not drop off radically. For example, for the case of 50 percent noise, *num_neurons* = 20 and *nlrn* 3750, the observed value of VR when only the three correct input variables were used was 42.8 percent. When an additional seven extraneous input variables were also included, the observed VR was 35.7 percent. The *time_per_epoch* for these two cases was 2.26 seconds for *ncp* = 3 and 8.23 seconds for *ncp* = 10. For noisy problems, extraneous variables are more problematic. For example, for the case of 90 percent noise, in Table D.1 it is seen that the average value of VR for *num_neurons* = 20 and *nlrn* = 7500 is 7.74 percent. Running the same case with *ncp* = 6 (i.e., three extraneous variables), the value of VR dropped off to 5.76 percent. For *ncp* = 10 (i.e., seven extraneous variables) VR was down to 1.02 percent.

NOTES

1. S. Haykin, *Neural Networks—A Comprehensive Foundation* (New York: Macmillan Publishing, 1994).

2. W. D. Patterson, *Artificial Neural Networks—Theory and Applications* (Englewood Cliffs, NJ: Prentice Hall, 1996).

3. M. Smith, *Neural Networks for Statistical Modeling* (New York: International Thompson Computer Press, 1996).

4. A. P. Refenes, *Neural Networks in the Capital Markets* (New York: John Wiley & Sons, 1995).

5. E. Azoff, *Neural Network Time Series Forecasting of Financial Markets* (John Wiley & Sons, 1994).

6. J. S. Zirilli, *Financial Prediction Using Neural Networks* (New York: International Thompson Publishing, 1996).

7. E. Gately, *Neural Networks for Financial Forecasting* (New York: John Wiley & Sons, 1996).

8. W. Dwinnell, "Modeling Methodology, 1, 2, and 3", *PC AI,* Nov/Dec 1997, Jan/Feb 1998, Mar/Apr 1998.

9. G. E. Hinton, "Deterministic Botzmann Machine Learning Performs Steepest Descent in Weight-Space", *Neural Computation* 1 (1989): 143–150.

10. I. Alexander and H. Morton, *An Introduction to Neural Computing* (New York: Chapman and Hall, 1990).

11. M. Hagen and M. Menhaj, "Training Feedforward Networks with Marquardt Algorithm", *IEEE Transactions on Neural Networks* 5(6) (1994).

12. M. Hagen, H. Demuth, and M. Beale, *Neural Network Design* (Boston, Mass: PWS Publishing Company, 1996).

13. T. Masters, *Advanced Algorithms for Neural Networks* (New York: John Wiley & Sons, 1995).

14. K. Levenberg, *Quart Applied Math* 2(1944): 164.

15. D. W. Marquardt, *J. Soc. Applied Math*, 11(1963): 431.

16. H. Demuth and M. Beale, *Neural Network Toolbox Users Guide* (for use with MATLAB), The Math Works, Inc., 1997.

17. J. J. Stephan, "Complexity of Learning", In *Mathematical Perspectives on Neural Networks* edited by P. Smolensky, M. Mozer, D. Rumelhart (Mahwah, NJ: Lawrence Erlbaum Associates Inc., 1996).

18. L. Blum and R. Ronald, "Training a 3-Node Neural Network is NP-Complete", *Neural Networks* 5(1) (1992): 117-127.

19. T. Masters, *Private Communication*, 1999.

20. M. Zacksenhouse, *A MATLAB Routine: FASTLM.M,* mermz@techunix.technion.ac.il, 1999.

APPENDIX E

A TEST FOR SIGNIFICANCE OF FRACTION_SAME_SIGN

In Section 5.1, the output parameter $FracSS$ (Fraction_Same_Sign) is mentioned as a measure of the "goodness" of a model. In Figure 5.2 we see that $FracSS$ is included as part of the output of the FKR program. One reason for including this parameter is that it is so easy to compute. It is also easy to interpret, and its significance can easily be tested. (The following test of significance is taken from my course notes from a graduate course that I teach called the *Design and Analysis of Experiments.*[1])

The definition of $FracSS$ is simply the fraction of predicted values of Y that have the same sign as the actual values of Y for the test data set. Using the symbol y for the predicted values, we can express this relationship as follows:

$$FracSS = \frac{\sum_{i=1}^{i <= ntst} (Y_i y_i > 0)}{ntst} \tag{E.1}$$

If the model has no predictive power, we would expect the value of $FracSS$ to be a function of p_1 (the fraction of positive values of Y) and p_2 (the fraction of positive values of y). Defining p as the expected value of $FracSS$ for the null hypothesis, we find that that relationship between p, p_1 and p_2 is as follows:

$$p = p_1 p_2 + (1 - p_1)(1 - p_2) = 1 - p_1 - p_2 + 2p_1 p_2 \tag{E.2}$$

For example, if the model has no predictive power and if the value of $p_1 = 0.5$, we would expect a value of $p = 0.5$ regardless of the value of p_2. If the value of $p_1 = 0.6$, we would expect a value of $p = 0.4 + 0.2\,p_2$.

The null hypothesis assumes that for each test point the probability of y_i and Y_i having the same sign is p. For $ntst$ test points, the actual value of n (the number of test points in which the predicted and actual values of Y are the same) would be distributed as a binomial distribution with a mean value of $ntst*p$ and a standard deviation according to Eq. (E.3)

$$\sigma = \sqrt{ntst * p * (1 - p)} \qquad \text{(E.3)}$$

For large values of $ntst$ (i.e., $ntst >> 1$), u approaches a normal distribution with mean 0 and $\sigma = 1$:

$$u = \frac{n - ntst * p}{\sigma} \qquad \text{(E.4)}$$

The test for significance of a given value of $FracSS$ can thus be stated as follows:

1. For the given value of $FracSS$ compute n (the actual number of "same_sign" test points): $n = FracSS * ntst$
2. Examine all the test points to determine the values of p_1 and p_2.
3. Compute p using Eq. (E.2).
4. Compute σ using Eq. (E.3).
5. Compute u using Eq. (E.4).
6. For a given confidence level test the value of u based on a normal distribution of mean 0 and $\sigma = 1$.

For example, assume a measured value of $FracSS = 0.58$, $ntst = 5000$, $p_1 = 0.52$ and $p_2 = 0.53$. Is this value of $FracSS$ significant using a confidence level $= 0.01$? The value of n for this example is $5000 * 0.58 = 2900$, the value for p is $1 - 0.52 - 0.53 + 2*0.52*0.53 = 0.5012$, the value for σ is $sqrt(5000*0.5012*0.4988) = 35.36$, and the value of u is $(2900 - 5000*0.5012)/35.36 = 394/35.36 = 11.14$. Since this value is far above the 1 percent confidence level (i.e.,

2.33) for a standard normal distribution, we can conclude that the value of *FracSS* of 0.58 is highly significant.

NOTES

1. J. Wolberg, *Course Notes, Design and Analysis of Experiments* (Unpublished) Technion-Israel Institute of Technology, 1999.

BIBLIOGRAPHY

I. Aleksander and H. Morton, *An Introduction to Neural Computing* (New York: Chapman and Hall, 1990).

H. Akaikhe, "Statistical Predictor Identification," *Ann. Inst. Stat. Math* 22 (1970): 203-217.

J. S. Armstrong, *Long Range Forecasting: From Crystal Ball to Computer, Second Edition* (New York: John Wiley & Sons, 1985).

D. Aronson, "Pattern Recognition Signal Filtering," *Journal of the Market Technicians' Association* (Spring 1991).

S. Arya, D. Mount, N. Netanyahu, and R. Silverman, "An Optimal Algorithm for Approximate Nearest Neighbor Searching," In *Symposium on Discrete Algorithms*, Chapter 63 (New York: Springer-Verlag, 1994)

E. Azoff, *Neural Network Time Series Forecasting of Financial Markets* (New York: John Wiley & Sons, 1994).

Y. Bard, *Nonlinear Parameter Estimation* (New York: Academic Press, 1974).

J. M. Bates and C. Granger, "The Combination of Forecasts," *Operations Research Quarterly* 20 (1969): 451-469.

R. J. Bauer, *Genetic Algorithms and Investment Strategies* (New York: John Wiley & Sons, 1994).

R. E. Bellman, *Adaptive Control Processes* (Princeton, NJ: Princeton University Press, 1961).

C. M. Bishop, *Neural Networks for Pattern Recognition* (New York: Oxford University Press, 1995).

L. Blum and R. Ronald, "Training a 3-Node Neural Network is NP-Complete" *Neural Networks* 5(1) (1992): 117-127.

G. E. P. Box, G. M. Jenkins, G. Reinsel, and G. Jenkins *Time Series Analysis: Forecasting and Control*, 3rd edition, (Englewood Cliffs, NJ: Prentice Hall, 1994).

T. S. Chandee and S. Kroll, *The New Technical Trader* (New York: John Wiley & Sons, 1994).

T. S. Chandee, *Beyond Technical Analysis* (New York: John Wiley & Sons, 1997).

D. Culler and J. P. Singh, *Parallel Computer Architecture: A Hardware/Software Approach* (San Francisco, CA: Morgan Kaufmann Publishers Inc., 1999).

W. W. Daniel, *Applied Nonparametic Statistics* (Boston: PWS Publishing, 1990).

H. Demuth and M. Beale, *Neural Network Toolbox Users Guide* (for use with MATLAB) (The Mathworks, Inc., 1997).

J. Durbin and G. S. Watson, "Testing for Serial Correlation in Least Squares Regression," *Biometrika* 58 (1971): 1-19.

W. Dwinnell, "Modeling Methodology, 1,2 and 3," *PC AI*, Nov/Dec 1997, Jan/Feb 1998, Mar/Apr 1998.

P. D. Feigin and J. R. Wolberg, "A Test of Significance for the Cells Method Screening in Nonparametric Regression" (Unpublished, 1999).

J. C. Francis, *Investment, Analysis, and Management* (New York: McGraw-Hill, 1980).

J. E. Freund, *Mathematical Statistics* (Englewood Cliffs, NJ: Prentice Hall, 1992).

P. Gans, *Data Fitting in the Chemical Sciences* (New York: John Wiley & Sons, 1992).

T. Gasser, H. G. Muller, W. Kohler, L. Molianari, and A. Prader, "Nonparametric Regression Analysis of Growth Curves," *Annals of Statistics* 12 (1984): 210-229.

E. Gately, *Neural Networks for Financial Forecasting* (New York: John Wiley & Sons, 1996).

M. Hagen and M. Manhaj, "Training Feedforward Networks with Marquardt Algorithm," *IEEE Transactions on Neural Networks* 5(6) (1994).

M. Hagen, H. Demuth, and M. Beale, *Neural Network Design* (Boston, Mass: PWS Publishing, 1996).

W. Hardle, *Applied Nonparametric Regression* (Cambridge, UK: Cambridge University Press, 1990).

S. Haykin, *Neural Networks—A Comprehensive Foundation* (New York: Macmillan Publishing, 1994).

G. E. Hinton, "Deterministic Botzmann Machine Learning Performs Steepest Decent in Weight-Space," *Neural Computation* 1 (1989): 143-150.

N. L. Johnson and S, Kotz, *Continuous Univariate Distributions* (Boston: Houghton Mifflin Co., 1970).

M. Jurik, *Computerized Trading* (New York: New York Institute of Finance, 1999).

P. J. Kaufman, *Trading Systems and Methods,* third edition (New York: John Wiley & Sons, 1998).

B. W. Kernighan and C. J. Van Wyk, *Timing Trials: Experiments with Scripting and User-Interface Languages,* http//kx.com/a/kl/examples/bell.k, 1997.

P. R. Krishnaiah and L. N. Kanal, *Classification, Pattern Recognition, and Reduction of Dimensionality,* Handbook of Statistics (North Holland, 1982).

K. Levenberg *Quart. Applied Math.* 2 1944:164.

S. Makridakis et al., *The Forecasting Accuracy of Major Time Series Methods* (New York: John Wiley & Sons, 1984).

B. B. Mandelbrot, "A Multifractal Walk Down Wall Street ," *Scientific American* (February 1999).

H. Markowitz, "Portfolio Selection," *Journal of Finance* (March 1952).

D. W. Marquardt *J. Soc. Ind. Applied Math* 11 (1963): 431.

T. Masters, *Practical Neural Network Recipies in C++* (San Diego, Calif: Academic Press, 1993).

T. Masters, *Advanced Algorithms for Neural Networks* (New York: John Wiley & Sons, 1995).

J. T. McClave and P. G. Benson, *Statistics for Business and Economics* (New York: Macmillan Publishing, 1994).

W. Mendenhall and T. Sincich, *Statistics for Engineering and Science,* 3rd ed. (New York: Macmillan Publishing, 1992).

J. P. O'Shaughnessy, *What Works on Wall Street* (New York: McGraw-Hill, 1997).

R. Pardo, *Design Testing, and Optimization of Trading Systems* (New York: John Wiley & Sons, 1999).

W. D. Patterson, *Artificial Neural Networks: Theory and Applications* (Englewood Cliffs, NJ: Prentice Hall, 1998)

G. F. Pfister, In *Search of Clusters: the Ongoing Battle in Lowly Parallel Computing* (Englewood Cliffs, NJ: Prentice Hall, 1998).

F. P. Preperata and M. I. Shamos, *Computational Geometery: An Introduction* (New York: Springer-Verlag, 1985).

M. B. Priestley, *Non-Linear and Non-Stationary Time Series Analysis* (New York: Harcourt, Brace & Co., 1988).

D. Pyle, *Data Preparation for Data Mining* (San Francisco, CA: Morgan Kauffman Publishers Inc., 1999).

M. A. Ruggiero, Jr., *Cybernetic Trading Strategies* (New York: John Wiley & Sons, 1997).

T. Subba Rao, *Developments in Time Series Analysis* (in honor of M. B. Priestley) (New York: Chapman and Hall, 1993)

A. P. Refenes, *Neural Networks in the Capital Markets* (New York: John Wiley & Sons, 1995)

H. Samet, *The Design and Analysis of Spatial Data Structures* (Reading, Mass.: Addison Wesley Longman, 1990).

J. Schwager, *Fundamental Analysis* (New York: John Wiley & Sons, 1995)

J. Schwager, *Schwager on Futures: Technical Analysis* (New York: John Wiley & Sons, 1996).

W. F. Sharpe, *Investments* (Englewood Cliffs, NJ: Prentice Hall, 1979)

S. Siegel and N. J. Castellan, *Nonparametric Statistics* (New York: McGraw-Hill, 1988)

M. Smith, *Neural Networks for Statistical Modeling* (New York: International Thomson Computer Press, 1996).

J. J. Stephan, "Complexity of Learning," In *Mathematical Perspectives on Neural Networks* edited by P. Smolensky, M. Mozer, D. Rumelhart (Mahwah, NJ: Lawrence Erlbaum Associates Inc., 1996).

H. Tong, *Threshold Models in Nonliner Time Series Analysis* (New York: Springer-Verlag, 1983).

A. Ullah and H. D. Vinod, "General Nonparametric Regression Estimation and testing in Econometrics," *Handbook of Statistics 11* (North Holland, 1993).

A. Wald and J. Wolfowitz, "Statistical Tests Based on Permulations of the Observations," *Ann. Math. Statistics* 15 (1984): 358-372.

A. S. Weigend and N. A. Gershenfeld: *Time Series Prediction: Forecasting the Future and Understanding the Past* (Reading, Mass: Addison Wesley Longman, 1994).

N. Weiner, *The Extrapolation, Interpolation, and Smoothing of Stationary Time Series with Engineering Applications* (New York: John Wiley & Sons, 1949).

S. M. Weiss and N. Indurkhya, *Predictive Data Mining: A Practical Guide* (San Francisco, CA: Morgan Kauffman Publishing Inc., 1998).

L. Williams, Long-Term Secrets to Short-Term Trading (New York: John Wiley & Sons, 1999).

J. Wolberg, *Prediction Analysis* (New York: Von Nostrand Reinhold, 1967).

J. Wolberg, Course Notes, Design and Analysis of Experiments, Technion-Israel Institute of Technology (Unpublished).

H. Wold, *A Study in the Analysis of Stationary Time Series* (Stockholm, Sweden: Almqvist and Wiksell, 1938).

M. Zacksenhouse, *A MATLAB Routine: FASTLM.M*, mermz@techunix.technion.ac.il, 1999.

J. S. Zirilli, *Financial Prediction Using Neural Networks* (New York: International Thompson Publishing, 1996).

INDEX